# Doing Educational Administration

*A Theory of Administrative Practice*

# Titles of Related Interest

Evers and Lakomski/Knowing Educational Administration: Contemporary Methodological Controversies in Educational Administration Research

Evers and Lakomski/Exploring Educational Administration: Coherentist Applications and Critical Debates

Hodgkinson/Administrative Philosophy: Values and Motivations in Administrative Life

MacPherson/Educative Accountability Policies for Educational Institutions

# Doing Educational Administration

*A Theory of Administrative Practice*

by

## COLIN W. EVERS

*The University of Hong Kong, Hong Kong, China*

and

## GABRIELE LAKOMSKI

*University of Melbourne, Australia*

2000

# PERGAMON

An imprint of Elsevier Science
Amsterdam – Lausanne – New York – Oxford – Shannon – Singapore – Tokyo

371.2
E93d

ELSEVIER SCIENCE Ltd
The Boulevard, Langford Lane
Kidlington, Oxford OX5 1GB, UK

First edition 2000

**Library of Congress Cataloging-in-Publication Data**
Evers, C.W.
   Doing educational administration: a theory of administrative practice / by Colin W. Evers, Gabriele Lakomski.
      p. cm.
   Includes indexes.
   ISBN: 0-08-043351-0
   1. School management and organization—Philosophy. I. Lakomski, Gabriele. II. Title.

LB2805 .E915 2000
371.2'001—dc21                                                                                         99–054250

**British Library Cataloguing in Publication Data**
A catalogue record from the British Library has been applied for.
ISBN 0-08-043351-0

∞ The paper used in this publication meets the requirements of ANSI/NISO Z39.48-1992 (Permanence of Paper).

# Dedication

It is most fitting that this book be dedicated to the memory of Don Willower, 1927 - 2000. We can never repay our debt to our Mentor and dear friend for the generous and gracious support he gave us professionally as well as personally. Let this book be a small token of our gratitude and thanks. And let it also be an indication of our intention to carry on in the naturalistic tradition in educational administration which he pioneered and represented so formidably. We think he would like that!

Colin Evers and Gabriele Lakomski
Hong Kong and Melbourne

# Contents

*Part III: Researching Practice*

# Preface

This is the third volume in a series of books in which we have attempted to develop a systematic approach to educational administration. Our aim in this book is to explain and apply recent developments in the theory of knowledge representation to a range of problems and issues concerning the practice of educational administration. In pursuing this aim, we build on two bodies of knowledge. The first is our earlier work showing how to adopt a scientific stance on administration that is neither some version of positivism nor an excursion into the realms of postmodernism. For us, the content and structure of theory in educational administration is shaped by appeals to the global excellence of such theory, to its overall coherence, including the requirement that it cohere with natural science. This perspective on knowledge justification is known as naturalistic coherentism.

The second body of scholarship to which we appeal is recent work in cognitive science that explores the contribution natural science can make to an understanding of how the brain actually represents and processes knowledge. Since about the mid-1980s, work on computer models of the brain's cognitive processes, particularly the construction of artificial neural networks (ANNs), has proceeded with impressive results. Since most of the problems associated with developing a theory of administrative practice end up being concerned, in one way or another, with difficulties over how to represent non-symbolic knowledge — knowing *how* rather than knowing *that*, to use the usual dichotomy — appeal to ANN models of knowledge representation, which do not require practitioners to possess symbolic accounts of their know-how, seemed to us to be the most promising way forward.

We therefore commence, in Part I of the book, with an introductory account of this research, though one illustrated with a number of examples of concern to educators. A general framework for understanding the representation of practical knowledge is proposed, first for individuals, and then for groups where the social distribution of knowledge is important. The position presented in this part is not to be seen as functioning as some basis for deriving recommendations about specific administrative practices. The specifics are simply not at hand for that purpose. Rather, in keeping with our epistemology, our theory of practice should be seen as functioning as a general constraint on the understanding and conduct of particular practices. More precisely, accounts, analysis and recommendations concerning particular practices need to cohere with the theory of practice. Just as we have been able to show that coherence with natural science works as a strong constraint on the formulation of theories of educational administration, so we show that theory of practice powerfully constrains options for understanding such matters as leadership, organizational design, decision-making, ethical conduct, and administrator training. This task, at any rate, is the burden of Part II.

We conclude, in Part III, with a coherentist overview of research methodology, explore some ways of extending it to cohere with our naturalistic account of practice, and offer a summing up of the main ways in which our naturalism has consequences for researching practice.

Naturalism has been, early and late, our overarching philosophical orientation. That is, we expect our general theory of educational administration in all its aspects and applications to cohere with and make use of the most relevant theories from natural science. The reason we adopt a naturalistic stance, rather than some other perspective, is not because we see natural science as providing some kind of foundation to knowledge. Rather, we conjecture that its adoption as a central feature of systematic theorizing about educational administration will lead to a more coherent, and hence more justified, view of these phenomena. To the extent that our extension of this perspective into theory of administrative practice is fruitful, our conjecture derives further warrant, making for a more complete and comprehensive account of educational administration.

Colin Evers and Gabriele Lakomski
The University of Hong Kong and University of Melbourne

# Acknowledgements

The development of the ideas in this book has been a gradual process. We have again followed our usual practice of presenting portions of work in progress to different audiences. Some chapters have had their origins in conference papers, while others were written specifically as book chapters for other editors, or as journal articles. In all cases we are very grateful for the critical comment and attention that this work, in its development, has received. Where prompted by helpful suggestions, and our own second thoughts, all previously published work has been revised, sometimes beyond recognition from its original version.

Portions of Chapter 1 were first presented as an Invited Address to the Administration Division of the *American Educational Research Association* in San Francisco, 1995, and subsequently published, in 1996, in the *Educational Administration Quarterly*. The last two thirds of the chapter are taken from material that appeared first in the *Journal of School Leadership*, in 1998, following an invitation from Donald Willower and the journal's Editorial Board, to address these themes.

An ancestor of Chapter 2, containing its more philosophical parts, was prompted by an invitation from David Aspin to contribute to his volume *Logical Empiricism and Post-Empiricism in Educational Discourse* (1997) in the Heinemann series on Philosophy of Education.

Chapter 3 was developed in response to an invitation from John Keeves for a chapter in *Issues in Educational Research* (1999). Some aspects of this chapter were originally discussed at an international conference at Umeå University, Sweden, by invitation of the Director of the Centre for Principal Development, Olof Johansson, in 1998, and appeared in the Conference Proceedings, *Exploring New Horizons in School Leadership* (1998).

Chapter 4 arose out of a presentation to the Inaugural Conference of the OISE/University of Toronto and UCEA *Centre for the Study of Values in Educational Leadership*, held in Toronto, in 1996. Some aspects of this chapter were also discussed at a Division A (Administration) sponsored symposium on values and leadership at the *American Educational Research Association* annual meeting in San Diego, in 1998. Further developments were published, at the invitation of Paul Begley and Pauline Leonard in their book, *The Values of Educational Administration: A Book of Readings* (1999).

Most of the work in Chapter 5 was first presented at the international conference at Umeå University's Centre for Principal Development (Sweden), and appeared in the Conference Proceedings (1998). Select aspects of this chapter were also discussed at an Organizational Theory SIG Session on *Organization Theory and Organization Learning* at the *American Educational Research Association* annual meeting in Montréal, Canada, in 1999.

The work presented in Chapter 6 was first developed for a seminar course for Masters degree students at the University of Hong Kong. Useful feedback from members of the class contributed greatly to making the material more accessible to a wider audience.

Portions of Chapter 7 were also first presented at the Inaugural Conference of the OISE/University of Toronto and UCEA *Centre for the Study of Values in Educational Leadership*, held in Toronto, in 1996. The work was subsequently developed, at the invitation of Paul Begley, with some appearing in his book *Values and Educational Leadership* (1999) and some appearing in a volume also co-edited by Pauline Leonard, *The Values of Educational Administration: A Book of Readings* (1999).

Chapter 8 is a considerably expanded and revised version of an article that first appeared, by invitation, in the *UCEA Review*, in the Fall of 1998. Some aspects of the chapter were discussed at the Annual National Conference of the *Australian Council for Educational Administration* in 1998.

An early version of Chapter 9 was presented, by invitation, at a conference on research methodology at the University of Newcastle, Australia, in 1996. It was later revised and has appeared in *Issues in Educational Research* (1999). Select aspects were discussed, by invitation, at *The David L. Clark National Graduate Student Research Seminar in Educational Administration and Policy*, sponsored by UCEA, AERA, Divisions A and L., and Corwin Press, at the *American Educational Research Association* annual meeting in Montréal in 1999.

Work in progress on Chapter 10 was presented at a staff seminar at the University of Hong Kong. Parts of the chapter have also appeared in *Issues in Educational Research* (1999).

We are grateful for all the advice and comments we received from audiences to whom this work was presented, and to particular individuals who read portions of the manuscript and gave generously of their wise counsel. In particular we would like to thank Nicholas Allix, David Aspin, Paul Begley, Olof Johansson, John Keeves, Pauline Leonard, Mark Mason, Viviane Robinson, and Kam-cheung Wong, for their time, attention, and support for this research.

For permission to reprint previously published material, we are grateful to various publishers. Full bibliographic details of all of our previous publications we have drawn on are given below:

EVERS C.W. (1997). Philosophy of education: A naturalistic perspective, in D.N. Aspin (ed.) *Logical Empiricism and Post-Empiricism in Educational Discourse*, (Johannesburg: Heinemann), pp. 167–181.

EVERS C.W. (1998). Decision-making, models of mind, and the new cognitive science, *Journal of School Leadership*, **8**(2), 94–108.

EVERS C.W. (1999). Complexity, context and ethical leadership, in P. Begley and P. Leonard (eds.) *The Values of Educational Administration*, (London: Falmer Press), pp. 70–82.

EVERS C.W. (1999). From foundations to coherence in educational research methodology, in J. P. Keeves and G. Lakomski (eds.) *Issues in Educational Research*, (Oxford: Pergamon Press), pp. 264–279.

EVERS C.W. AND LAKOMSKI G. (1996) Science in educational administration: a postpositivist conception, *Education Administration Quarterly*, **32**(4), 379–402.

LAKOMSKI, G. (1998). Training administrators in the wild: A naturalistic perspective, *UCEA Review* , **XXXIX**(3), Fall.

LAKOMSKI, G. (1998). Leadership, distributed cognition and the learning organization, in O. Johansson and L. Lindberg (eds.) *Exploring New Horizons in School Leadership* — Conference Proceedings Skrifter från Centrum för Skolledarutveckling, Umeå universitet, Umeå, pp. 98–112.

LAKOMSKI, G. (1999). Symbol processing, situated action, and social cognition: Implications for educational research and methodology, in J.P. Keeves and G. Lakomski (eds.) *Issues in Educational Research*, (Oxford: Pergamon Press), pp. 279–300.

LAKOMSKI, G. (1999). Against leadership: concept without a cause, in P. Begley and P. Leonard (eds.) *The Values of Educational Administration: A Book of Readings*, (London: Falmer Press), pp. 36–50.

LAKOMSKI G. AND EVERS C.W. (1999) Values, socially distributed cognition, and organizational practice, in P. Begley (ed.) *Values and Educational Leadership*, (Albany, N.Y.: SUNY Press), pp. 165–182.

Finally, for valuable assistance in the preparation of this manuscript, we would like to thank Cindy Wu of the Centre for Educational Leadership, and Sophia Yam and Samuel Lau of the Support Unit for Educational Research and Development, at The University of Hong Kong.

# PART I

# Cognition and Practice

# 1

# Theory, Mind and the New Cognitive Science

The aim of this book is to use new developments in cognitive science as a basis for understanding how to theorize about practice in general and about the practice of educational administration in particular. The three chapters comprising Part I deal with more general issues about cognition and practice. Ideas in these chapters are not so explicit that they can be used to derive specific recommendations for the conduct of educational administration. That is rather too ambitious for our purposes. However, we do show that these ideas can act as constraints in the sense that some specifics cohere more readily with them than with others. Since we also regard appeals to coherence as the most central feature of justifying knowledge, we claim that our theory of practice has some purchase on the problems of formulating, choosing and implementing courses of action.

This book is part of a wider research project in which we signal the importance of theory of knowledge, or epistemology, for educational administration. In earlier work (e.g. Evers and Lakomski 1991, 1996) our focus was on the significance of epistemology for developing a theory of administration and adjudicating among rival theories. For sustained theorizing in educational administration, we advocated naturalistic coherentism as a background epistemology. Our present task is to use this epistemology, suitably augmented with recent findings from cognitive science, to develop a theory of practice in a way that readily meshes with our position on administrative theory.

In this chapter we begin with a brief overview of this earlier work and then move on to describe some ideas in recent cognitive science that can be used to develop an account of how knowledge of practice should be represented. Ultimately, it is our naturalistic stance on knowledge representation that does most of the work.

## Epistemology and Theory in Educational Administration

The content and structure of theories of educational administration have been much influenced by theories of knowledge and theories of human cognition (Evers and

Lakomski 1991). These influences are significant, affecting not just broader issues concerning the nature of theory, relations between theory and practice, or the nature of organization, but also quite specific issues such as leadership, organizational design, or administrator training (Evers and Lakomski 1996). Many of the characteristic features of behavioural science approaches to educational administration are due to an underlying logical empiricist theory of knowledge, while features associated with more recent rival views reflect corresponding alternative epistemologies. This development has been made possible by two philosophical shifts.

First, a series of powerful arguments had been mounted against logical empiricism, which provided the epistemological framework in which science was largely understood. These arguments, advanced during the same period of the Theory Movement's rise to prominence, by Quine (1951, 1960), Sellars (1963), Hanson (1958) and Feyerabend (1963) demonstrated the complexity of any relationship between a theory and its supporting empirical evidence. For example, observation reports always employ some theoretical vocabulary, and so are not theory free or theory neutral. Since observations are theory laden the possibility is raised of using a theory to reject an observation, rather than the other way around. As a consequence, because measurement procedures are expressed in some theoretical vocabulary, the possibility of giving operational definitions is also compromised. Furthermore, the same finite set of observations can function to confirm any number of different theories, because theories are always underdetermined by evidence. We then have the problem of determining precisely which theory is being confirmed. Disconfirmation, on the other hand, is problematical because of the complexity of test situations, which raises uncertainties over which parts of a theory need to be revised (Evers and Lakomski 1991, pp. 19–45). One result of these difficulties is that empirical evidence alone never seems sufficient for rationally preferring one theory to another. There seem to be no adequate foundations for knowledge sufficient to adjudicate among competing theories.

The second development was provided by Thomas Kuhn's account of the growth of scientific knowledge, first published in 1962. From studies in the history of science Kuhn concluded that for large-scale theories, or paradigms, it is the theory itself which determines the nature of the epistemic relationship between evidence and theory; that is, what counts as justification becomes paradigm specific. With no corresponding account of epistemic relations *between* theories possible, shifts between paradigms entered the domain of sociological knowledge. One consequence of these philosophical arguments has been a rejection of scientific objectivity and a subsequent vigorous defense of subjectivity, particularly as advocated in the influential work of Greenfield (1975, 1979, 1980, 1986; Greenfield and Ribbins 1993).

Much the same kind of critique was occurring in other branches of educational studies. For example, the paradigms perspective is now the norm in educational studies, especially in educational research methodology, and has in turn further abetted the flourishing of alternative viewpoints in administration with their associated epistemologies. Values perspectives, critical theory, and cultural theory all constrain the scope of science, while that most recent development, postmodernism, in following Rorty's (1980, 1991, 1992) advice of surrendering on epistemology altogether, reduces science to just another narrative.

When reviewing the epistemological assumptions of the main theoretical positions in the field in our book *Knowing Educational Administration*, we argued that the whole framework of debate between science and its critics was mistaken. The biggest mistake ironically involves an uncritical acceptance of positivist accounts of the nature of science — and then arguing that since positivism is plainly inadequate for various purposes, so too is science. Our strategy has been to detach science from positivism and argue the merits of a postpositivist view of science and its justification; one drawing on a coherentist and naturalistic view of knowledge. We agree that foundational patterns of justification are mistaken and that empirical evidence is never sufficient for rational theory choice. However, some theories are plainly of more value than others when it comes to building bridges, fixing cars, inventing computers, promoting learning, running an organization, and avoiding (or promoting) social tyranny. We argued that these, and other examples of the utility of learning, ubiquitous in nature, are best explained by positing the operation of criteria of theory choice *additional* to empirical adequacy. These additional, or *superempirical*, virtues of theory include consistency, simplicity, comprehensiveness, fecundity, and explanatory unity, and taken together with empirical adequacy, function in what we call a *coherence theory of justification* (see Churchland 1985; BonJour 1985; Williams 1977; and Lycan 1988). We think that other epistemologies, including versions of postmodernism, make tacit use of coherence justification in advancing their claims, so we argue that our view of justification is often common, or touchstone, theory in methodological debates. Finally, our theory of knowledge draws attention to the requirement that all epistemologies need to cohere with accounts of how humans learn, or acquire knowledge. Since we think that the best theories of learning come from natural science accounts of human information processing in the brain, rather than the *a priori*, or commonsense, accounts more typical of philosophical invention, our coherentism coheres with natural science. Our effort to develop a new science of educational administration aims to produce an administrative theory which is part of the most coherent global theory that coheres with natural science.

Although this is an enormous task, which is best approached as an agenda for more particular research items, one big advantage is that it should not result in a loss of knowledge. Where logical empiricism's view of science omitted much that was of value to administration, and where paradigms views are apt to diminish much that is of value in science, coherentism avoids discounting knowledge which is more warranted than the methodology doing the discounting. In this book, we hope to extend coherentism's reach to the domain of practical knowledge.

Recent major developments in educational administration may reasonably be characterized as a succession of attempts to challenge one or more of the assumptions of traditional science of administration.

Although there is room for controversy over our chronology and the theories we have selected, the central features of these challenges remain much the same. Thus,

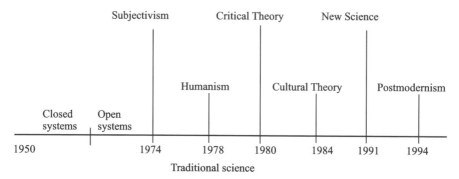

**Figure 1.1** Historical development of theoretical positions in educational administration.

if we identify behavioral science with a continuous tradition that runs from what is known as the Theory Movement of the 1950s, and exemplified in the early writings of Halpin (1957) and Griffiths (1959), through to the ideas expressed in successive editions of Hoy and Miskel's (1996) influential text, four logical empiricist assumptions of behavioural science stand out:

1.  A theory is a hypothetico-deductive structure where less general and singular claims, including expected observations, are deduced from more general claims (and defining context).
2.  Theories are justified by meeting certain conditions of empirical testability. If the deduced states of affairs are actually observed, the theory is confirmed. Otherwise it is disconfirmed. Justifying a theory is a matter of showing it has more confirmations and fewer disconfirmations than its rivals do.
3.  All the theoretical terms of a theory must be able to be given operational definitions, that is, admit of some defining measurement procedure.
4.  Scientific theories of educational administration exclude substantive ethical claims.

In observing the development of theory in the field, it is clear that Thomas Greenfield's work, from his 1974 IIP address through to his last papers, takes issue with all four assumptions. From his reading of Kuhn and Feyerabend, he knew that the justificationist assumption was wrong. Classical confirmation theory is compromised by the fact that observations are always embedded in theory. Moreover, since the same set of observations can confirm any number of distinct theories, it is problematic knowing exactly which is being confirmed. Theory ladenness also affects disconfirmation because it is no longer clear whether there are problems with the theory or whether they lie with the disconfirming empirical evidence. And the complexity of deductive relations within a theory makes for uncertainty in establishing the case for a direct empirical hit on any individual hypothesis. For Greenfield, the blurring of theory and observation undercut any attempt to establish a workable notion of objectivity in natural science. Together with his acceptance of

Weber's view of the place of subjectivity in social science — that social realities are constituted by meanings, interpretations, intentions, and understandings — the hypothetico-deductive construal of theories loses its point if the quest is for understanding. There are no theory free measurement procedures to halt a regress in operationalism, and values are as much a part of human subjectivity as the rest of the framework that makes for meaningful action and association (Greenfield and Ribbins 1993). His talk of multiple realities, of people inhabiting different worlds, and of paradigms, provided a major stimulus towards creating the by now commonplace *theoretical pluralism* in the field.

Another influential theorist, Christopher Hodgkinson's (1978, 1983, 1991, 1996), disputes principally the fourth assumption. Consistent with behavioural science, he readily accepts an unbridgeable gap between fact and value, and that science is about facts not values. Contrary to the usual inference, however, Hodgkinson claims that administration is inextricably value laden, and hence cannot be developed or adequately characterized as a science; rather it is a humanism. Also equating science with positivism, critical theorists, especially those influenced by the early writings of Habermas (e.g. 1972), argued for a broader source of knowledge for administrative theory, a source which acknowledged the relevance of communicative knowledge and emancipatory knowledge in human affairs. (Bates 1983, 1995; Foster 1980, 1986) Unfettered by the constraints thought to attend scientific knowledge, this more generous approach also runs against all four assumptions of logical empiricism. Cultural theorists typically move beyond these assumptions by adopting an irreducibly subjective account of social science and championing its distinctiveness from a supposedly limited natural science. Consistent with Greenfield's position, the social world is assumed to be constructed partly out of our representations of it, how these representations are interpreted by others, and how we in turn interpret the interpretations of others (the double hermeneutic). As a result, very little of the apparatus of behavioural science can be relocated within this 'human invention' perspective (Sergiovanni and Corbally 1984).

While maintaining an opposition to important elements of logical empiricist epistemology, and hence its theory of objectivity, most administrative theorists who construe their work as lying within a postpositivist matrix of alternative paradigms still maintain a representational interpretation of theory, albeit attenuated by anti-foundationalist arguments against science. But if foundationalism is thought to be the only way of defending justification, these arguments will eventually spill over into an attack on the very possibility of epistemology. This indeed has happened with postmodernism as championed by Rorty and Bernstein (1983, 1992). Systematic expressions of postmodernism are still very new in educational administration. (See, for example, Maxcy 1994, and the Special Issue of the *Educational Administration Quarterly*, August, 1998, on Postmodernism.) Nevertheless, there is no question that in their more radical subjectivist denial of the four assumptions of logical empiricism, they promise a continuation of the theoretical debate and innovation that has been characteristic of educational administration during the last twenty five years.

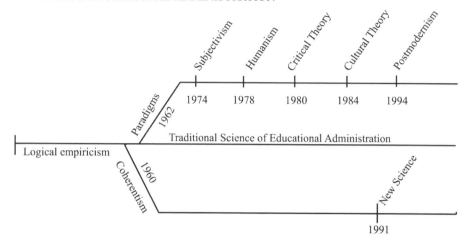

**Figure 1.2** Epistemological structure of historical development of theories in educational administration.

## Naturalistic Coherentism: The New Science

While Figure 1.1 gives a rough chronology of the main epistemological dissenters from science in educational administration, it omits an important structural feature concerning the range of possible options within theory of knowledge. Most criticisms of logical empiricism that lie within the paradigms tradition imply that justification is either seriously weakened, or lapses entirely, because of difficulties with empirical adequacy as a criterion of theory choice. However, a general discounting of justification leaves problematical the question of why some theories rather than others are vastly more successful in enabling us to solve problems, fulfil our expectations, and anticipate the course of future experience; why, for example, we can foresee the behaviour of others with some measure of success, or catch a plane with a better than chance expectation of arriving at the correct destination.

Our adoption of a coherence theory of justification, namely one which draws on more criteria than just empirical adequacy, is motivated precisely because we accept both arguments against empirical foundationalism (advanced by subjectivists and postmodernists) *and* arguments for the capacity of humans to extract useful patterns from an evidently non-random flux of experience. Figure 1.2 therefore gives a more realistic portrayal of the epistemological options driving alternative perspectives within administrative theory.

Our research program is about exploring the ways in which a coherentist epistemology, and what it requires, places constraints on the content and structure of substantive theories in educational administration. Because epistemology is not an *a priori* exercise, but makes assumptions about what humans can know and how learning occurs, an important requirement of our coherentism is a *theory of cognition*. In order to determine how useful patterns are extracted from experience, we opt for a view of cognition informed by our best natural science, an approach to epistemology pioneered by Quine in his (1960) book *Word and Object*. (See also Quine 1969). Our choice of natural science is, of course, quite fallible; science is not here functioning as an epistemic foundation. Rather, the choice will be vindicated to the extent to which coherence advantages accrue to the surrounding global theory in which this account of cognition is embedded.

For us, the coherentist alternatives to the four assumptions of logical empiricism discussed earlier are as follows:

1.  Concerning structure, coherentism implies a shift towards holism. Theories of administration are simply a topic specific part of a continuous web of belief that comprises the global theory we are all working up from infancy onwards. There is some weak structure within the web, but it is imposed by degree of entrenchment. Statements at the perimeter, perhaps singular observation reports, are those which would be most readily revised in the light of experience. Those at the centre, in our view logic, mathematics, and branches of physics, function as major organizing features of the web and are least revisable, unless doing so makes for substantial gains in simplicity, or some overall gain in the coherence of the global theory. A science of administration for us therefore is an administrative theory that is part of a coherent web of belief whose most central organizing claims concern logic, mathematics and the natural sciences.

2.  Justification is construed as epistemically progressive learning. A fallible web gives rise, through ongoing social practice, to theoretically motivated expectations which either match or mismatch theory laden experience. Revisions to the web, or a portion of it such as administrative theory, are justified if they lead to greater overall coherence.

3.  Theoretical terms derive much (but not all) of their meaning, not from their proximity to some particular observation procedure, but by virtue of the role they play in the conceptual scheme of a theory. What appear to be operational definitions can be seen in a more relaxed way as proposed measurement procedures selected for appropriateness, perhaps for reasons of empirical precision, according to an already existing set of meanings of terms.

4.  The value ladenness of experience is of a piece with the theory ladenness of observation. Values are learned and justified, therefore, along with the rest of our theory of the world, through the same process of coherent adjustment in the light of experience. Their apparent distance from observation is a reflection of theoreticity, of the positioning of values more to the centre of a web, than any methodological distinction between facts and values. Because of its subject matter, we take it as uncontroversial that values are embedded throughout educational administration theory.

As learning is presumed to take place across theoretical perspectives, we expect these considerations will similarly cut across all the so-called paradigms and intrude into non-representational postmodernist domains as well. However, the key point here is that they extend into the issue of how we understand practical knowledge.

Our central strategy is to naturalize epistemology itself. That is, our naturalistic coherentism should be reflexive and thus cohere with scientific accounts of how knowledge is acquired and represented by knowers. Since, from an epistemological perspective, practices need to be learned, and can enjoy success and failure in a way analogous to theories, we also embrace the prospect of naturalizing theory, looking to the result as a way of developing a unified account of theory and practice. In the remainder of this chapter, we take a first look at what is involved in offering a naturalistic view of cognition, by identifying three significant background models of mind. The aim is to run contrary to the usual custom of trying to portray practice in cognitively laden linguistic terms, by examining how a characteristically cognitive activity, namely decision-making, might be construed in non-linguistic ways. By moving away from an emphasis on the symbolic as a basis for representing knowledge, we raise the possibility of a common representational framework for both practical and theoretical knowledge.

Of the three models of mind that have informed methodological approaches to cognition, the first to be considered is the 'absent mind' approach, embodying assumptions that have informed the behaviorist tradition of research. Secondly, developing out of perceived weaknesses in behaviorism, at least as an account of cognitively oriented behaviors, is the 'functionalist mind' approach, which underwrites traditional cognitivism and its manifestations in program writing artificial intelligence. Finally, we examine the 'material mind', an approach that seeks to limit functionally adequate accounts of cognition by using the requirement that they be more biologically realistic. That is, theories about human reasoning are to be constrained by the demand that they cohere with theories concerning the representation and dynamics of knowledge in the brain. We use the expression 'the new cognitive science' to refer broadly to those efforts at theorizing about cognition that attempt to meet this constraint.

## Cognition in the Absent Mind

The intense focus on administrative behavior among researchers in educational administration during the 1950s and 1960s flowed from a methodological commitment to what is loosely referred to as positivism, but what might be labeled with more technical accuracy as logical empiricism (Evers and Lakomski 1991, pp. 46–75.) On this view, all elements of an administrative theory were both meaningful and justifiable if they could be shown to correspond to some specific range of sensory experience. Since the inner cognitive and affective workings of humans cannot be directly observed, the field became populated with theories of administrative behavior in general and leadership behavior, interpersonal behavior, and decision-making behavior, among the particulars (Halpin 1966).

In his classic, *Science and Human Behavior* (1953, p. 35), Skinner sums up the case for the irrelevance of mind as follows:

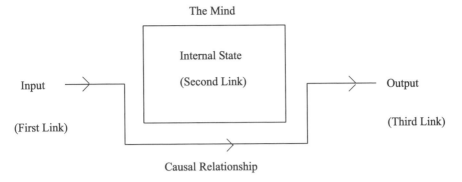

Figure 1.3 Bypassing the mind.

> The objection to inner states is not that they do not exist, but that they are not relevant in a functional analysis. We cannot account for the behavior of any system while staying wholly inside it; eventually we must turn to forces operating on the organism from without. Unless there is a weak spot in our causal chain so that the second link is not lawfully determined by the first, or the third by the second, then the first and third links must be lawfully related.

What he means by this argument can be seen more clearly from Figure 1.3.

If there are no causal gaps, then the posited set of internal states comprising the second link ($S2$) will be some function of the external input states comprising the first link ($S1$). Thus $S2 = F(S1)$. By the same reasoning, the output states ($S3$) will be a function of the internal states, so that: $S3 = f(S2)$. But the no gaps condition will then ensure that there is another, perhaps more complex, function that can be used to describe the connection between $S1$ and $S3$ directly: $S3 = F' S1)$. Therefore, if we wish fully to explain a repertoire of decision behaviors a person exhibits under certain conditions, we need only look at the network of relevant causes impinging on that person under those conditions. Since any variation in output caused by the mind will be reflected in observable variations in inputs to the mind, we can let so-called invisible 'mind variables' drop out and just deal with complex relations of inputs to outputs.

What were thought to be the relevant input variables to decision outcomes? Halpin's (1957) brilliant paper 'A Paradigm for Research on Administrator Behavior' gives something of the flavor of this kind of research. In addition to recording attributes and characteristics of administrators such as 'age, intelligence, academic training, and experience as a teacher or as a school administrator' (Halpin 1957, p. 59) it is crucial that the administrator's perception of the organization's task be taken into account. This requires that decision and other problems, *as perceived*, should 'be stated exclusively in behavioral terms' (Halpin 1957, p. 46). Fortunately, a person's spoken and written comments, their analyses and interpretations of organizational life, letters, telephone calls, the scheduling of meetings, all count as behaviors.

Decision outcomes also need to be formulated behaviorally, preferably using terms similar to those found in the vocabulary for describing inputs. Given appropriately tabulated behavioral data sets, the chief aim of the 'Paradigm' as a research tool is to establish evidence for correlations among the inputs and outputs. With the addition of some agreed rating for the merits of outcomes, behavioral criteria for effective decision-making can be posited that might be used to sustain programs of administrator training.

Despite vast amounts of effort having been put into this research program over the years, it contains a number of fundamental difficulties. First, while there may be no causal gaps, it is entirely another matter to identify precisely which input stimuli in a given situation are regularly associated with particular cognitively oriented output behaviors. Chomsky (1959) makes this point with devastating effect in his review of Skinner's (1957) *Verbal Behavior*, when he considers a person responding to a painting with either the words 'Dutch' or 'Clashes with the wallpaper' or 'Never saw it before' or 'Beautiful' or anything else that might come into our mind. Second, the fact of creativity compromises the business of finding correlations. Third, strictly speaking there is an arbitrarily large number of ways in which environmental events can be classified into relevant inputs. However, to make it all work in a plausible way, a background heuristic seems to be tacitly employed that selects inputs into those that might be attended to, or those of interest, or those that were being thought about — a heuristic, in other words, that appeals to 'inner states'. Fourth, the basic criteria for appraising decision-making *strategies,* have a global quality. For example, using the term 'rational' to describe strategies for the efficient use of means to bring about ends effectively, is partly a comment on a complex structure of interrelated behaviors that for these purposes will not usefully reduce to a consideration of separable inputs and outputs. And finally, at this level of cognitive complexity, the evaluation of decision-making and other cognitive behaviors is characteristically done on *representations* of these phenomena — invariably symbolic representations — which are assumed to be in some sense *in the mind* guiding administrator behavior.

For these reasons, it is easy to see why dealing directly with symbolic representations themselves became the center of interest in understanding decision-making.

## Cognition in the Functionalist Mind

Traditional cognitivism goes beyond behaviorism in that it is willing to posit an inner structure of so-called second link variables to account for the coordination of inputs and outputs. Indeed, its most impressive contributions have concerned attempts to theorize the nature of cognition in terms of symbolic representations of inner structures. The bulk of Herbert Simon's work on administrative decision-making and bounded rationality constitutes the most sophisticated body of research within this tradition in administrative studies. Central to Simon's contribution is what Newell and Simon (1976, pp. 109–111) call the Physical-Symbol System Hypothesis, which states that 'A physical-symbol system has the necessary and sufficient means for general intelligent action', where such a system is taken to be 'a machine that produces through time an evolving collection of symbol structures'.

What this bold claim means is that cognitive tasks like planning, reasoning, deliberating, calculating, decision-making, evaluating, learning, knowing, believing, and thinking, can be understood roughly in terms of a three stage process. First, devise a symbolic way of representing objects and events in the world that can be physically encoded into a machine the way (for example) a computer program can be. Next, arrange for the machine to operate on its stored symbols according to a set of rules which are also stored, to produce further symbolic structures some of which may be termed outputs. Finally, connect the output in such a way that it is evidenced by behavior. (See Copeland 1993, p. 80). In a computer, output produced behavior might be printing, or screen displays.

On this model of cognition it does not matter what physical substance is used to code symbols. It can be vacuum tubes, magnetic relays, transistors, silicon chips, or neurons. The distinguishing properties of mind reside in the software, not the hardware. Influenced by Logical Empiricism, traditional cognitivism placed very weak constraints on accounts of mind. Since, as in the case of behaviorism, it was only inputs and outputs that were deemed methodologically observable, the principal empirical constraint on conjectures about the workings of mind was simply that conjectures had to be adequate as a function for transforming observed inputs into required outputs. This test is known as functional adequacy and is what lies behind functionalist models of mind.

To see how it works, consider a calculator inside a black box. The inputs are pressings of numbered keys and operation keys, the outputs are numbers on a liquid crystal visual display. Now suppose I enter some number and the operation 'divided by 5' and the calculator correctly computes the result every time. What theory can be legitimately conjectured to explain this modest piece of machine intelligence? Is the symbolic rule, physically encoded in the calculator, of the following kind: 'double the input number and divide by 10?' Or is it: 'check each digit from the left for divisibility by 5, perform the division and carry any remainder on to the next digit and repeat?' Well, the point is that these two different 'minds' are both functionally equivalent. They both give identical outputs to the same inputs.

Because of these modest constraints on conjectures about mind, the task of understanding decision-making shifts away from an emphasis on psychology and towards an emphasis on the formal properties of symbolic representations. Armed with the physical-symbol system hypothesis, criteria for intelligent decision-making are thus more appropriately defined over symbol systems. Many of these systems are quite arcane: optimization techniques, critical path analysis, maximization of expected utility under conditions of uncertainty, actuarial risk analysis, and a host of others. However, the point can be made with a very simple formulation. We need a representation of some desirable goal, $G$, perhaps expressed by the string '$G$ is a desirable goal'. Likewise for some means, $M$, reckoned as the most efficient and effective way of achieving $G$. Finally, we need to put these premises together, with any other premises thought necessary validly to deduce the decision to implement *M* to bring about *G*. (See Evers and Lakomski 1996, pp. 2–3). Now on the functionalist approach a decision looks like a species of argument, to be adjudicated as good if it involves valid reasoning about true premises. Of course, the experience of implementation may teach us that some premises were actually false, or that through ignorance we had omitted from the argument an important premise, or that the form of argument was too simplified. Moreover, the fact that our cognitive powers are quite limited, suggests that these mistakes will be common. Nevertheless, even lapses are reckoned in term of the representational features of a decision.

Intelligent action, seen from the perspective of humans possessing an intelligent program, like a computer program, has obvious consequences for administrator training. Instead of having one's behaviors shaped by a schedule of contingencies, the focus is on loading up a good program: in short, teaching theory, expressed as symbolic/linguistic strings for representing administrative situations, for operating on those strings, and for deducing decision outcomes. For the mind of a good decision-maker is one that can be said to contain a good program.

## Problems with the Functionalist Mind

Notwithstanding its widespread appeal, there are severe problems with functionalism, though these often go unnoticed because the medium of analysis is also symbolic/linguistic. Consider the old joke that for philosophers, humans are always rational, and for psychologists, they never are. A large body of evidence, built up over many studies by researchers such as Tversky and Kahneman (1981) shows quite clearly that ordinary human reasoning under a range of humdrum conditions is systematically flawed, at least in certain areas. We are subject to cognitive illusion. So regardless of how elegant and useful mental programs for decision-making may be, they are most probably idealizations of how we think rather than useful guides to the psychology of cognition.

Model builders respond to this criticism by distinguishing between normative decision theory, which is designed to tell us how we *ought* to make decisions, and descriptive decision theory, which is all about how we *do in fact* make decisions. Our 'natural' program contains a few bugs whose effects can be ameliorated by some normatively adequate re-programming.

This response sounds well because it is assumed that there is not a lot of difference between the two; they are just separated by a bit of fine tuning. Nor should the assumption be surprising, since the physical symbol system hypothesis was defined on a normatively selected subset of human behaviors, namely *intelligent* behavior. But what happens if our best descriptively adequate accounts of human cognition diverge so much from the implied symbolic mechanisms of the functionalist approach underwriting vast amounts of the normative tradition, that the whole notion of the mind as a program is compromised?

In our view, the divergence spills over to create a major gap between normative theory and administrative practice, causing problems with both the implementation of the theory and the teaching of it as a guide to practice. Here is why. First, suppose that the input coding problem can be solved and we have some language-like representation of administrative life that is reasonably comprehensive: there exist sentences for all the things that matter. Yet with just a few simple logical operations on a bare handful of these sentences, a huge number of deductions is possible, more than could ever be performed in a lifetime. (For $n$ sentences, the number of combinations is 2 raised to the power of 2 raised to the power of $n$. That is, 2 to the $n$th power in turn used as a power of 2.) Clearly, one way to limit the problem of combinatorial explosion is to opt for some heuristic limitation on deduction relative to, say, context or topic in order to frame the use of calculation. Unfortunately, beyond artificial, or toy universes, the 'frame problem' has proved intractable, thus limiting the utility of appeals to context in modeling real world reasoning.

Second, for many decisions we possess nothing that even remotely resembles a symbolic formulation. Take the decision on when to change gear in a car with a manual transmission. The engine sounds as though it's laboring a bit, but we're nearly at the top of the hill and there are no sharp turns immediately beyond. On the other hand, there are three passengers in the car. The resulting decision seems to be a response to a set of mostly perceptually presented soft constraints, rather than an unconscious deduction from some tacitly formulated propositions. We know this because we cannot say in words precisely what the engine sounds like when it's ready for a change of gear. A surprisingly large number of our judgments have this quality. When philosophers distinguish between 'knowing how' and 'knowing that', or between skills and propositional knowledge, the whole arena of skills, or non-propositional knowledge seems, almost by definition, to fail to have a program representation.

A failure of the theory of representation, when it comes to explaining the dynamics of skilled performance, ramifies extensively across the domain of practice. For while there may be an abundance of true propositions about teaching, leadership, surgery, swimming, acting, and decision-making, in all but the very simplest cases of performance the nature of the judgments involved resist capture by language, and rules for operating on language. That is why, in these domains, we promote learning by doing.

Third, and following on from the second point, the functionalist programming approach fails to take into account the hybrid nature of cognition, and in particular, how the symbolic and non-symbolic domains interact in the processes of human judgment. And for most professional skilled performance, intelligent action does take as its inputs mixtures or combinations of sentential strings, visual scenes, sounds, and smells — in fact whole ensembles of variously represented data. Because, from the perspective of the symbol system hypothesis, it is difficult to include the non-symbolic, the prospects for a comprehensive theory of intelligent practice are, in our view, to be found in recently emerging theories of how the brain represents and processes knowledge.

## The Material Mind

Methodologically, one of the ironies concerning traditional program-style cognitivism, was that while it placed no limitations on the construction of representations, save that they mapped observed inputs onto observed outputs, it was remarkably conservative in imagining what else might be going on inside the heads of people. Part of this conservatism was a result of the paucity of good theory on how thinking, reasoning, making decisions, or even something as ubiquitous as learning, could be accounted for in terms of brain processes. Moreover, earlier attempts to develop mathematical models of ensembles of neurons as information processors, were seen to be severely limited owing largely to an influential critique by Minsky and Papert (1969). Most of the blame for failure to move beyond functionally adequate language-like characterizations of thought, however, can be sheeted home to the multiple realisability argument. This asserts that since intelligence can be manifested in a variety of physical objects, locations and conditions, the analysis of intelligence can be conducted without the analysis of brains.

Whatever the merits of this argument (and in Chapter 2 we argue that they are slight) the last dozen years have seen a massive expansion of work on neural models of cognition leading to the recognition of a new development in cognitive science, what we call *the new cognitive science*. In what follows we want to discuss some of this work and sketch out some of its implications for understanding educational decision-making in particular and cognition in general.

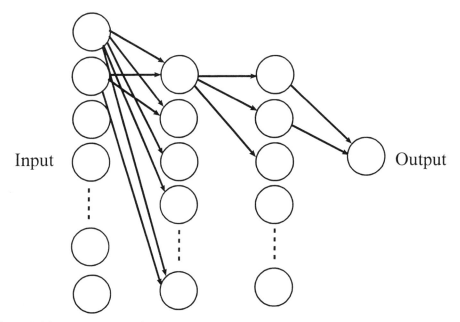

**Figure 1.4** A 20 x 10 x 10 x 1 four-layer neural network.

## Neural Networks: Some Examples

There are now many different models of how neurons process information. Some have been developed by engineers with only minimal attention being given to capturing the biological properties of neurons. These are often created for specific purposes, such as research on voice recognition by telephone companies, or postcode recognition by the post office. Others, for example those developed by Stephen Grossberg (1988) and his associates, are more attentive to biological detail because they seek to explain *human* cognitive phenomena. We shall not worry too much about this distinction here since the radically different nature of most brain models from their traditional functionalist rivals can be illustrated even with less realistic, and somewhat simplified, versions. We shall focus on one that has many engineering applications: something known as learning through backpropagation. Its main features can be seen in Figure 1.4.

In order to understand the various element of this network, how it learns, how it represents knowledge, and how it makes decisions, consider the kind of practical decision-making problem that it was designed to solve.

In the emergency department of a large hospital, when patients present with the symptom of acute chest pain, it is the job of physicians to make a diagnosis. In particular, they need to distinguish between patients who are having acute myocardial infarction (heart attack) and those who are not. Making a correct diagnosis will require physicians to take into account a range of evidence on the patient's record: age, sex, shortness of breath, hypertension, etc. Two sorts of errors are possible — turning patients away who are suffering infarction, and admitting patients who are not. The task is one that calls for a high level of trained professional judgment, adjudicating a complex and often conflicting body of evidence.

In an experiment to see if diagnostic accuracy could be improved, William Baxt (1990) developed and trained the neural network in Figure 1.4. This is a network that contains four layers: an input layer of 20 units, or nodes, two hidden layers of 10 units each, and an output layer of 1 unit. Each unit of the input layer corresponds to one of 20 positions in an input vector assigned numerical values associated with each of 20 symptoms entered onto the patient's record. Some of these symptoms are assigned simply a 1 or 0, indicating presence or absence of a condition, such as shortness of breath. Others are assigned finer grained numerical values, corresponding to age or blood pressure. These 20 numbers, arranged in a column vector, are fed into the matching nodes of the input layer. At the other end of the network, the one element output vector associated with the one unit output layer, can take either of two values: a 1 for presence or a 0 for absence of infarction. The aim is to train the network to produce the correct diagnostic output for each patient's input vector.

For calculating outputs, the network has a further design feature. Each node is connected to every other node in the adjacent layer, and for each of these connections between nodes there is an initially assigned number called a *weight* which is used to multiply the signal being transmitted from one node to the next. Nodes or units after the first layer need to sum all the incoming weighted signals and convert them to an output signal for the next layer. Since this output signal is the level of activity that node expresses after receiving its weighted inputs, it is computed by a function known as an *activation function* (Bechtel and Abrahamsen 1991, p. 40). Standard neural network software packages contain options for many different activation functions, with some being good approximations for the threshold firing properties of real neurons.

The last design feature of the network is a learning rule. Here is how learning, which in this network is known as supervised learning, occurs. The input vector initiates signals in the input layer which are transmitted down the connections to be weighted, summed, transformed by an activation function and sent along to the next layer and subjected to the same process. For each input vector the net computes an output vector. Because the numerical value of the weights is initially set randomly, the output would be correct only by chance. Supervised learning occurs by checking the net's output against the known correct output. Any difference between the two is error. A learning rule is a mathematical formula for computing the contribution each weight makes to producing error, and then changing weights by a computed amount in order to minimize that error.

Baxt (1990) trained his network to make diagnostic decisions as follows. His total data set consisted of 356 patients: 236 known not to have infarction, 120 with infarction. He made a random selection of half the patients from each group, to create a training set. The training set consisted of 178 input-output pairs. The entire training set was fed through the network again and again until the learning rule had adjusted all the weights to the point where the net produced a correct output vector for each patient record input vector. After learning correctly to identify the heart attack patterning within patient test data, the net was then fed the other half of the data on which it had not been trained. As Baxt (1990, p. 480) tells it:

> The network correctly identified 92% of the patients with acute myocardial infarction and 96% of the patients without infarction. When all patients with the electrocardiographic evidence of infarction were removed from the cohort, the network correctly identified 80% of the patients with infarction. This is substantially better than the performance reported for either physicians or any other analytical approach.

Under constant pressure of feedback (or backpropagation as it is called in this network design) the net had locked on to the pattern present in the data. The net had learned from experience.

Precisely where does the network's knowledge of true heart attack symptoms reside? Where is it physically located? The answer is that it is distributed across the geometry of the net within its configuration of weights and their values. (Biologically, the weights represent connection strengths at synaptic junctions between neurons.) Notice that this knowledge is not represented in some language-like rule for classifying patient data. Unlike traditional expert systems, which might solve the diagnostic decision problem by coding in a whole series of sentences of the 'if...

then...' form, derived from interviews with physicians examining the same data, there are no sentence formulations. The same set of distributed weights contains highly accurate diagnostic knowledge for each of the 356 patients.

While it might be thought that this is a net that works well only because the input data are not sentences or symbolic strings, notice that the concept of an input vector is very wide. Even sentences can be coded as input vectors. Indeed the human eye does it automatically, transforming sentence graphemes, or tokens, into arrays of spiking frequencies transmitted along the optic nerves, for specialized processing by ensembles of neurons located in the language centres of the cortex. A fascinating example of training a neural network on language data is NETtalk, which takes vector codings of words, seven letters at a time, as inputs and gives vector codings for phonemes as outputs. (See Churchland 1989, pp. 168–181, for an introductory discussion). Over several hours of feedback driven supervised learning, NETtalk learned to pronounce written English text, simply by a process of minimizing error between the net's output speech and the target speech data for the written text.

By a suitable vector coding of images, nets have been trained to recognize faces, and one face recognition network called EMPATH was trained to associate patterns of facial expression from inputted photographs, with named emotional states such as 'astonishment, delight, pleasure, relaxation, sleepiness, boredom, misery and anger' (Churchland 1995, p. 126). It performed quite well on the positive emotions, but fairly poorly on negative ones, ironically mirroring the more successful result profile of human subjects on the same task.

## Some Implications for Understanding Educational Decision-Making

Although research into artificial neural networks (ANNs) is fairly new, we think that some general lessons can be learned from the literature of the new cognitive science. The first is that most likely, humans possess powerful non-symbolic distributed representations of practical skills and the knowledge which underlies much expert judgment. A theory of human competence that demands symbolic expression of this knowledge will produce misleading results as to what a person 'really' knows. A person certainly requires an appropriately trained motor neural map to win a major golf tournament, but it is doubtful if their golf swing, or short game, depends on their capacity to describe that skill in words.

Second, much valuable learning occurs simply through the experience of doing something, and responding to the many sources of feedback that occur during and after that act of doing. (Indeed, it is arguable that, in evolutionary terms, the human cognitive system developed partly out of resulting cognitive demands initiated by coordinated motor performance). Even for cognitively high level tasks, such as copying files on a computer, our knowledge of these is most likely 'in the hands'. That is, we can probably work our computers quite effectively for familiar tasks, but be unable to provide correct written or spoken descriptions of each necessary step we take.

Third, because it is better to see reasoning as pattern processing rather than sentence crunching, the nature of judgment has a different analysis. When a school principal makes a decision, say, to admit a child with a particular disability into that school, instead of seeing the result as a deduction from a set of unarticulated premises, it is more plausible to see the input information as triggering a *prototype* of a successfully integrated child, a prototype that has been built up by experience. When an artificial network, such as the one for medical diagnosis is analyzed, what is found is that after training, the inputs that are classified as instances of infarction produce a characteristic pattern of activation across the hidden layers, while the non-infarction data produce a different characteristic pattern of activation. It is this pattern that may be regarded as the mind's 'prototype'. Even for simple objects like chairs, or tables, the mind does not classify according to necessary and sufficient properties. There are none. Classification is probably done by a neural equivalent of 'similarity' to some prototype, or experientially based average, for chairs and tables.

Fourth, unlike language, vector coding can permit the simultaneous inputting of both complexity of environment and continuous, fine grained, distinctions. More importantly, however, many of the natural inputs to real life decision problems are multi-modal, coming via a range of sensory modalities. Neural network models of cognition, in being able to account for these data, hold the promise of allowing us to understand the importance of a wider set of decision considerations that influence decision-makers. And simply acknowledging this point helps to remind us that it is a legitimate part of the scope of research into decision-making of the new cognitive science.

Fifth, it is useful to think of linguistic/symbolic formulations of knowledge as ways of *compressing* experience into a representation. Now these compression algorithms, or summaries, of experience will be most realistic when what they describe is relatively context invariant. Hence, mathematics, and mathematical formulations of situations, are most valuable in physics, where context is of minimal importance. Similarly, a symbolic theory formulation of leadership, or of decision-making, or of teaching, or of surgery, viewed as a compression algorithm will be of diminished value as a representation of, and a guide to, practice where contextual factors dominate. But fine grained distributed representations of knowledge are particularly good at accounting for contextual factors. For example, all the information concerning the 356 decisions made by Baxt's medical diagnostic model was contained in no more than 310 numerical weights. As a result, there is value in organizing training in complex, context dependent tasks as a form of *simulation* for people in possession of such vast neural resources. There has been a tendency to see simulation as an alternative to expensive learning by doing. We are suggesting that the motivation should be not to avoid expense, but to preserve realism of learning task.

Sixth, we want to draw attention to the fact that traditional functionalist inspired models of thought are weak on modeling thinking as a dynamic process that occurs in real time. Thought, conceived in terms of events in some abstract decision space, fails to capture the embeddedness of actors in a physical world of immediate or delayed consequences that can be a source of feedback, further learning, and the context for more decision-making. We would go further. In emphasizing the material mind, we emphasize its continuity with the body, and with the complex structure of material needs and wants that contributes to the development of values and that partly determines the nature of problems and acceptable solutions.

Finally, in the chapters that follow, it is the notion of a material mind that will be developed and its consequences for practice explored. It needs to be said that the question of the theoretical adequacy of particular brain models for understanding cognition is still very much the subject of active investigation within this approach to cognitive science, although on many particular, practical problems, they have had their successes. (For a technical introduction to some of this research see Hassoun 1995). Nevertheless, we think there is enough promising detail in such emerging disciplines as cognitive neurobiology, to encourage methodological and substantive moves beyond the now dominant functionalist approach to knowledge and its representation, and towards an examination of the constraints a more naturalistic, scientific approach imposes. It is here that we think the most fruitful insights for understanding skilled practice will be found.

## References

Bates R.J. (1983). *Educational Administration and the Management of Knowledge*. (Geelong: Deakin University Press).
Bates R.J. (1995). Critical theory of educational administration, in C.W. Evers and J.D. Chapman (eds.) *Educational Administration: An Australian Perspective*. (Sydney: Allen and Unwin).
Baxt W.G. (1990). Use of an artificial neural network for data analysis in clinical decision-making: the diagnosis of acute coronary occlusion, *Neural Computation*, **2**, pp. 4809–489.
Bechtel W. and Abrahamsen A. (1991). *Connectionism and the Mind*. (Oxford: Blackwell).
Bernstein R.J. (1983). *Beyond Objectivism and Relativism*. (Philadelphia: University of Pennsylvania Press).
Bernstein R.J. (1992). *The New Constellation: The Ethical-Political Horizons of Modernity/Postmodernity*. (Cambridge, MA: MIT Press).
BonJour L. (1985). *The Structure of Empirical Knowledge*. (Cambridge, MA: Harvard University Press).
Chomsky N. (1959). *A Review of B. F. Skinner's Verbal Behavior, Language*, **35**, pp. 26–58.
Churchland P.M. (1985). The ontological status of observables: In praise of superempirical virtues, in Churchland P. M. and Hooker C. A. (eds.) *Images of Science*. (Chicago: University of Chicago Press).
Churchland P.M. (1989). *A Neurocomputational Perspective*. (Cambridge, MA: MIT Press).
Churchland P.M. (1995). *The Engine of Reason, The Seat of the Soul*. (Cambridge, MA: MIT Press).
Copeland J. (1993). *Artificial Intelligence: A Philosophical Introduction*. (Oxford: Blackwell).
Evers C.W. and Lakomski G (1991). *Knowing Educational Administration*. (Oxford: Pergamon).
Evers C.W. and Lakomski G. (1996). *Exploring Educational Administration*. (Oxford: Pergamon).
Feyerabend P.K. (1963). How to be a good empiricist, in Baumriu B. (ed.) *Philosophy of Science. The Delaware Seminar*. (Newark: University of Delaware Press).
Foster W. (1980). Administration and the crisis in legitimacy: a review of Habermassan thought, *Harvard Educational Review*, **50**(4), pp. 496–505.
Foster W. (1986). *Paradigms and Promises*. (Buffalo, Prometheus Press).
Greenfield T.B. (1975). Theory about Organization: A new perspective and its implications for schools, in T.B. Greenfield and P. Ribbins (1993) (eds.) *Greenfield on Educational Administration: Towards a Humane Science*, pp. 1–25 (IIP 1974 paper, reprinted). (London: Routledge).
Greenfield T.B. (1979). Ideas versus data: how can the data speak for themselves?, in G. L. Immegart and W. L. Boyd (eds.) *Problem-Finding in Educational Administration*. (Lexington: Lexington Books).
Greenfield T.B. (1980). The man who comes back through the door in the wall: discovering truth, discovering self, discovering organizations, *Educational Administration Quarterly*, **16**(3), pp.26–59.
Greenfield T.B. (1986). The decline and fall of science in educational administration, *Interchange*, **17**(2), pp. 57–80.
Greenfield T.B. and Ribbins P. (1993) (eds.). *Greenfield on Educational Administration: Towards a Humane Science*. (London: Routledge).

Griffiths D.E. (1959). *Administrative Theory*. (New York: Appleton-Century-Crofts).

Grossberg S. (ed.) (1988). *Neural Networks and Natural Intelligence*. (Cambridge, MA: MIT Press).

Habermas J. (1972). *Knowledge and Human Interests*. (London: Heinermann).

Halpin A.W. (1957). A paradigm for research on administrator behavior, in Campbell R.F. and Gregg R.T. (eds.) (1957) *Administrative Behavior in Education*. (New York: Harper). Cited as reprinted in Halpin (1966).

Halpin A.W. (1966). *Theory and Research in Administration*. (New York: Macmillan).

Hanson N.R. (1958). *Patterns of Discovery*. (Cambridge: Cambridge University Press).

Hassoun M.H. (1995). *Fundamentals of Artificial Neural Networks*. (Cambridge, MA: MIT Press).

Hodgkinson C. (1978). *Towards a Philosophy of Administration*. (Oxford: Blackwell).

Hodgkinson C. (1983). *The Philosophy of Leadership*. (Oxford: Blackwell).

Hodgkinson C. (1991). *Educational Leadership: The Moral Art*. (Albany: SUNY Press).

Hodgkinson C. (1996). *Administration Philosophy*. (Oxford: Pergamon Press).

Hoy W.K. and Miskel C.G. (1996). *Educational Administration: Theory, Research and Practice*. (New York: McGraw-Hill, 5th Edition).

Lycan W.G. (1988). *Judgement and Justification*. (Cambridge: Cambridge University Press).

Maxcy S.J. (ed.) (1994). *Postmodern School Leadership*. Meeting the crisis in educational administration. (Westport, Connecticut, London: Praeger).

Minsky M. and Papert S. (1969). *Perceptrons*. (Cambridge, MA: MIT Press).

Newell A. and Simon H.A. (1976). Computer science as empirical enquiry: symbols and search. Cited as reprinted in Boden M.A. (ed.) *The Philosophy of Artificial Intelligence*. (Oxford: Oxford University Press).

Quine W.V. (1951). Two dogmas of empiricism, *Philosophical Review*, **60**, pp. 209–43.

Quine W.V. (1960). *Word and Object*. (Cambridge, MA: MIT Press).

Quine W.V. (1969). Epistemology naturalised in W. V. Quine *Ontological Relativity and Other Essays*. (New York: Columbia University Press).

Rorty R. (1980). *Philosophy and the Mirror of Nature*. (Princeton, NJ: Princeton University Press).

Rorty R. (1991). Objectivity, relativism, and truth, *Philosophical Papers*, Volume 1. (Cambridge: Cambridge University Press).

Rorty R. (1992). We anti-representationalists, *Radical Philosophy*, **60**, **Spring**, pp. 409–42.

Sellars W. (1963). *Science, Perception, and Reality*. (New York: Routledge and Kegan Paul).

Sergiovanni T.J. and Corbally J.E. (eds.) *Leadership and Organizational Culture*. (Chicago: University of Illinois Press).

Skinner B.F. (1953). *Science and Human Behavior*. (New York: Macmillan).

Skinner B.F. (1957). *Verbal Behavior*. (New York: Appleton-Century-Crofts).

Tversky A. and Kahneman D. (1981). The framing of decisions and the psychology of choice, *Science*, **211**, pp. 453–458.

Williams M. (1977). *Groundless Belief*. (Oxford: Blackwell).

# 2

# Representing Knowledge of Practice

The purpose of this chapter is to present our view of how practical knowledge should be understood. We do this by first considering some of the most influential traditions that have attempted to make sense of knowledge of practice, and second, by setting our own account in the context of features of our broader view of administrative knowledge, and of what we are calling the new cognitive science.

The conventional approach to practice as best exemplified in the work of Simon (1976, pp. 257–278)) or Hoy and Miskel (1996, pp. 266–271) can be captured by the following formula:

(1)   Theory + goals => action.

In this familiar model, 'theory' means propositional, or sententially expressed knowledge, while 'goals', also sententially expressed, are purported to lie outside of theory. In combination, however, they lead to action. Theory guides the accomplishment of goals through the medium of action. Action, thus construed, is intelligent and purposeful. Related versions of this model include: 'means + ends => action', 'policy + operations => implementation', and; 'strategic planning + operations => action'.

Several assumptions about knowledge lie behind this conventional approach. First, a relatively static view of knowledge is presumed, where the relevant theory comes before action. The model also relies heavily on that theory being correct, since it is directly antecedent to some kind of action. Finally, the model tends to be insensitive to the context dependent nature of useful knowledge, although this is more a reflection on how difficult it is to achieve specific relevant theory in the absence of an option for making adjustments in the light of consequences.

Since this conception of practical knowledge places largely unrealistic epistemic demands on practitioners, there has been considerable interest in drawing on epistemologies that capture the fluid, fallible and context dependent nature of knowledge for action. Two overlapping views have been prominent in educational studies. The first we call the 'Growth of Knowledge' tradition which includes Dewey's pragmatism. Its basic innovation is to link a recognition of the fallibility of all knowledge with a proposed mechanism for learning, or knowledge improvement. Acting on theory to realize goals then occurs within the additional constraint of maintaining knowledge acquisition. The key epistemic strategy is to provide a feedback loop from the consequences of action to the original theory that informed that action. Roughly speaking, Dewey's (1916, pp. 333–339; 1938, pp. 85–88) formula was:

(2)   Situated problem + need to solve it => experimental attempt at resolution.

Actually, this formula is to be seen as initiating a cycle that continues with new problems arising out of attempts at resolution. Additionally, the cycle is maintained by shifting circumstances that cause situated problems constantly to change. While Dewey regularly employed biological metaphors that might suggest that the cycle would come to a halt if there were no individually 'felt' need to solve a problem, it should be noted that problems can have an objective structure. That is, a problem can be thought of as capable of generating a felt need for whoever is in the situated circumstances.

Another approach, widely discussed in the 1960s and continuing to have support, is the model for the growth of scientific knowledge proposed by Karl Popper (1959, 1963). Popper adopts the following formulation to summarize his view of how scientific knowledge grows:

(3)   $P1 => TT => EE => P2$

where $P1$ is the initial (theoretically motivated) problem to which we respond with a tentative theory, $TT$, that embodies a trial solution. It is subsequently tested, or criticized in an effort to eliminate errors, $EE$, a process that gives rise to a new problem, $P2$. (Popper 1969) Because it is fallible, scientific knowledge needs to be conceived in terms of a rigorous scientific practice that embodies this schema in order to promote its growth. And for both Dewey and Popper, knowledge is assumed to be located in the theories that grow out of attempts to solve problems.

In the case of the 'Growth of Knowledge' tradition, it is still feasible to see a separation between knowledge as located in theories and the epistemic practices that lead to the improvement of that knowledge. Within a second approach, known broadly as the 'Praxis' tradition, there is a tendency to collapse this distinction or at least to minimize it. The classic formulation can be found in Book Six of Aristotle's *Nicomachean Ethics*, where praxis is defined as action guided by practical reason. Indeed, what distinguishes a piece of behaviour from an action is precisely the role of reason. An adequate description of some action would not only make reference to some aspect of the reasoning that led to it, but the action would be a 'logical outcome' of that practical reasoning since, for Aristotle, reason has causal powers.

Resonating with this Aristotelian framework is Paulo Freire's (1972) characterization of education as a form of political action, widely discussed in the 1970s. The schema he adopts for summarizing his praxis view of knowledge, namely,

(4)    Praxis = reflection + action,

actually incorporates action as a constitutive part of what it means to have knowledge of practice. (The social relations of learning that Freire inveighed against under the label of 'banking education' where students sat passively and had their minds filled up with knowledge 'deposits', did not inculcate a kind of knowledge that failed to articulate with action. Rather they led to inadequate action of the sort that made for social and political powerlessness.)

Perhaps the most widely discussed and applied version of knowledge as praxis in recent times is that contained in the work of Donald Schön (1983, 1987a) on the reflective practitioner. Although sounding Deweyan in its focus on learning by doing, it also has significant Aristotelean elements in encouraging educators to see student behaviour in terms of the reasons students might conceivably have for their action. Calling for a new 'epistemology of practice', Schön (1987b), develops the notion of 'knowing-in-action' to describe the kind of taxonomy of conceptual structure that results from attending to the way ideas and concepts cohere in practice.

To illustrate the difference between traditional school knowledge and his new view, Schön takes an example from Vygotsky. Classifying together from a larger set, the objects hammer, hatchet, and saw as tools, is a form of school knowledge, which is formal and categorical. But grouping together hatchet, saw, and wood, taken from the same larger set, reflects the coherence of practice where experience is primarily of the use of the hatchet and the saw to produce firewood for cooking and warmth. Presumably, practical reason provides the conceptual glue for holding these practice focussed clusters of concepts together. Despite its origins in practice, if 'knowledge-in-action' sounds too static a concept, the dynamics of practical conceptual change are captured by the term 'reflection-in-action', which is meant to signify the improvisation and spontaneity manifested by reasoning agents in response to the shifting and sometimes unpredictable circumstances of practice. For Schön, the activities of teaching and administration are best theorized as forms of reflection-in-action.

As is evident from the discussion in the previous chapter, we have reservations not just about conventional ways of characterizing knowledge of practice, but also about influential formulations of both the Growth of Knowledge and the Praxis traditions. These reservations are of two types. First, there is an excessive reliance on linguistic formulations for purposes of representing practical knowledge. We do not deny that a great deal of intelligent, purposeful, human behaviour can be made to fit a predictive model that supposes people behave *as if* they are manipulating sentence tokens in their heads in a rational, rule-like way. Our point is that even if such a model were empirically adequate, fitting all relevant behaviour, we know enough about the causal dynamics of human cognition to know that it is more likely that some other model is being equally confirmed by this empirical evidence. Now if this latter model coheres with a more extensive range of good theory than its sentential token crunching rival, then we have coherentist epistemological grounds for favouring its *explanations*. Such is the claim of our naturalism.

Our second reservation concerns the fact that where these sentential models recognize limits to their representation of knowledge of practice — and Schön, for example, is certainly willing to do this — they say almost nothing about what else practical knowledge might be; its nature, acquisition, dynamics and representation. Part of the reason this is not always perceived as a serious omission is because explanatory standards have been shaped by the methodological requirements of either the 'absent mind', or the 'functionalist mind'. On the 'absent mind' approach, manifest within talk of 'outcomes based training' or 'training for competencies', one simply identifies practical knowledge with the behaviours exhibited, thus leaving nothing epistemic to be explained. And while the functionalist approach supposes some causally intervening mode of representation between behaviours and their appropriately described circumstances for occurring, it can be agnostic about the details. Hence, talk of *reasons* can usefully link the domains of circumstance and behaviour without producing any methodological demand to specify precisely how reasons manage to produce behaviour, or why they might produce the behaviours they do.

## Levels of Explanation

Before saying more about options for a naturalistic representation of practical knowledge, we want to deal with an influential defence of functionalism proposed by Herbert Simon (1995) in his advocacy of analysis at the symbolic level rather than at the neuronal level. Simon (1995, pp. 26–27) begins by observing that many physical systems are hierarchically organized; for example, organs are made up of cells, which are made up of molecules, which are made up of atoms, and so on. Or societies are made up of organizations, subgroups, and individuals. Hierarchy, in this sense, is made possible because of a feature of some complex systems known as near-decomposability. Roughly speaking, what this means is that the relative stability of assemblies composed of smaller components at a lower level is due to an averaging effect. Acquiring the skill of catching a football in flight does not require a calculation of the trajectories of each of its various component molecules. By dint of averaging, the molecules exhibit enough global integrity for football talk to figure as useful explanatory vocabulary within the art and science of football training and practice. The same goes for countless other middle sized material objects whose hierarchically lower components have been gerrymandered into artifacts of social and cultural significance. In the domestic environment of the kitchen, explanation and prediction that trades in a theoretical vocabulary pitched at the level of coffee cups, toasters, plates and cutlery is invariably more than sufficient for the tasks at hand.

And so, the argument runs, is recourse to the symbolic in accounting for human thought. The symbolic captures, at a higher level, the aggregate behaviour of assemblies of neurons. Moreover, since intelligent thought and action can be realized by material objects with very different physical features, the domain of the symbolic is, as a worthwhile explanatory resource, relatively autonomous. Known in the literature as the 'multiple realizability argument', Simon (1995, p. 33) puts it thus:

> Neurons and chips have little in common beyond the fact that they can both think, and that the latter can be programmed to think in humanoid fashion. This fact demonstrates the existence of a symbolic (software) level of theory above the hardware or neuronal level.

The main problem with this kind of argument is that it abstracts too severely from the physical context in which rational thought and action occur. (See Churchland, P. S. 1995, pp. 105–107.) Suppose, for example, we want to know why neurons and chips can think but toasters cannot. At some point in the explanation we need to give a detailed account of why certain physical configurations sustain behaviours describable at the symbolic level and others do not. Since such an account would amount to explaining a higher level of aggregated behaviour in terms of properties applicable to a lower level of component detail, the apparent separateness of higher levels of system functioning and their associated modes of conceptualization is relative to the kind of system behaviour in need of explanation.

Another example can illustrate this point. Consider the problem of meshing an account of intelligent thought with an account of how such thought might be learned. Despite the phenomenon of learning being multiply realizable, the best recent research on human learning does not offer explanations at the generic level. Instead, effort is expended on trying to explain how human brains acquire and process information. The current research literature tends to be dominated by conceptual and mathematical models that attempt to capture the causal detail of brain functioning most responsible for learning. In short, 'bottom-up' explanations in this domain are beginning to prevail over those that are 'top-down'. The traffic here is one way. The more the bottom-up strategy succeeds in delivering fine grained knowledge of any of the phenomena that might be reasonably expected to mesh with higher level ones, the more the explanatory apparatus associated with these higher level phenomena might be expected to mesh conceptually with concepts that prevail at the lower level. That is, it is a mistake to see thinking, on the one hand, and learning to think, on the other, as falling into two distinct explanatory domains.

When it comes to characterizing rational action, the demand for interlevel meshing would seem to be compelling. For example, for all the independent high level rationality that can be marshaled in favour of bidding at an auction, what does it profit a person if their hand, thoroughly enmeshed in an antecedent causal field, fails to rise to the occasion? And yet, again, the demand on explanatory resources is relative to what is supposed to require explanation. So for Aristotle, Simon, and a host of others, *X*'s hand went up *because X* desired to make a bid and raising one's hand in these circumstances is the rational thing to do. End of story. Except as Adrian Cussins (1992) points out, there is the little matter of explaining what appears to be a miraculous coincidence between what is accounted for by higher level concepts such as the symbolic, and lower level concepts such as the neuronal and physiological. To quote Cussins' (1992, p. 197) example:

> Every time you venture on to the road and obey the convention to drive on the left, think to yourselves: Isn't it a miracle that the events in the nervous system which control your arm on the steering-wheel cause the wheel to be in just the place required to satisfy your intention to drive on the left?

It is a tribute to the pervasiveness of top-down, higher level conceptual structures in theorizing about rational action, or practice, that the miraculous coincidence problem passes largely unnoticed. It is as though there is no significant methodological difference between the action of a disembodied soul and that of a physically constituted person! Nevertheless, because rational action is embodied action we feel entitled to press for a coherent rendering of the different levels, and a more physical account of rationality in the context of human practice.

## Natural Representation

Part of the argument against Simon alluded to a mistake in running together empirical adequacy and explanatory adequacy. The difference is easy to see. Grandma's bunion, animated as it is by discriminating animal spirits, may compete just fine with the national weather bureau in predicting the nuances of local weather. But when it comes to explaining these phenomena, the advantages of interlevel articulation between the atmospheric and the molecular, by way of the disciplines of thermodynamics and fluid dynamics, become persuasive. While weather talk is high level, explanation runs deep, being motivated by coherence considerations such as unity, simplicity, and comprehensiveness.

We can often see the need for going deeper, or down a level or two, when empirical adequacy fails, or when its lapses cannot be accommodated smoothly. Failures of learning in school are one area where this sometimes occurs. Obviously, folk-theoretic belief-desire theory can go some of the way in explaining lapses: thus, a child may not learn some task because it desires attention and believes that behaving in a manner not conducive to learning will bring attention. Countless discipline policies, with their graded sanctions against defined misconduct, presume the efficacy of belief-desire theory. Nevertheless, the stubbornness of some lapses has prompted a division of labour that is now well established in education. Just as mainstream education appears to deal with the higher level ideal of developing reason, Special Education has a methodological focus on causes, on the study of the brain and its perceptual systems. It is almost as though the detail of neuroscientific accounts of learning is simply irrelevant outside the context of Special Education. Yet neural processing underwrites educational performance in both contexts. A coherent account of knowledge, teaching and learning, essential for any systematic theory of education, would place a premium on fitting together these divergent aspects of educational experience into a smooth, comprehensive framework.

As we saw in the previous chapter, a shift to the causal, micro-cognitive aspects of knowledge and its acquisition, can be motivated partly by perceived gaps in the more familiar linguistic/symbolic tradition. And these gaps are often focussed around areas of practice, where skilled performance occurs in the absence of significant symbolic strings that can be identified as representing a practitioner's knowledge. Gaps disrupt the smooth flow of high level explanation, prompting more detailed, lower level explanations. Once attention shifts from the functionalist mind to the material mind, neural network accounts of the micro-structure of human knowledge loom large. One challenge, to which we give consideration, is avoiding the emergence of a complementary set of failures of the micro-causal to explain the symbolic. Hence our willingness to see the natural processing of physical symbol tokens as a species of pattern processing. What follows reprises and further elaborates the ideas introduced earlier.

Despite an accumulation of knowledge about how neurons create, transmit and process signals, only in the last fifteen years or so have good mathematical models been widely available to capture this behaviour. (Rumelhart and McClelland 1986) These models, and there are many different types, can be run on computers to test hypotheses about how ensembles of neurons learn and how they store what they have learnt. A considerable amount of commercially available software is now available for interested researchers (e.g. McClelland and Rumelhart 1988; Caudill and Butler 1992; Ward Systems Group 1993; Plunkett and Elman 1997).

In the case of artificial neural networks (ANNs) that learn through feedback from experience in meeting a goal or target, it is useful to distinguish two types of network architecture. The first, of which the Baxt model discussed in the previous chapter is an example, is given for three layers in Figure 2.1.

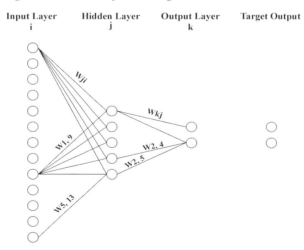

**Figure 2.1** A three-layer feedforward net with 13 input nodes, 5 hidden nodes and 2 output nodes. The target output nodes are not part of the net's design. Weights, Wji and Wkj, are adjusted by backpropagation of error to minimise their contribution to the difference between the output layer and target output. Not all connections are shown.

This network consists of three layers of nodes, or artificial neurons. The first is an input layer, the last an output layer, and the one in the middle is called a hidden layer. Each node in earlier layers is connected to every node in the next layer. As a signal from an input node is transmitted to the next layer, it is multiplied by a weight, usually a number between +1 and –1. At first these weights, which represent the synaptic junctions between neurons, are usually set at random, but they are gradually adjusted as the network learns some task. In a feedforward, backpropagation network, the input signal moves in one direction through the network to the output layer. Learning occurs when the net's output converges on some target output. A mismatch indicates an error that has been caused by the numerical value of the weights. An adjustment to weights, in proportion to their contribution to the error, is then transmitted back through the net in a process known as error correction by backpropagation. As input and target output pairs of patterns, or vectors, are presented over and over again, in an appropriately designed network the output vector will almost always converge on the target output. The net has then learned all the pairs in the data set. The knowledge that has been acquired is distributed across the whole network, residing in the configuration and value of all the weights.

Because of their parallel architecture, and the form in which information is presented and processed, neural networks are best regarded as pattern processors. For example, feedforward networks have been trained to identify types of music, the past tense of English verbs, the validity of simple argument forms in the propositional calculus, and to distinguish rocks from mines. (Churchland, P.M. 1995, pp. 57–96) They are being increasingly designed and trained to recognize more complex patterns, or regularities in language, which has generated considerable controversy as success in this application begins to threaten the dominant tradition that language is rule based. (For an introductory discussion, see Bechtel and Abrahamsen 1991, pp. 176–204.) Feedforward networks are very good at recognizing prototypical *things*, such as the correct number from among endless variations in handwriting. However, many characteristic cognitive tasks involve recognizing patterns in *processes*, that is, in sequences of events extended over time. Recognizing causes, for example, requires recognizing instances of prototypical processes. To model the kind of short term memory required for identifying patterns among temporally extended trajectories, we need to use *recurrent* networks. Their basic architectural features are captured in Figure 2.2.

Recurrent networks operate in much the same way as feedforward networks except that they have the additional design feature of feeding back outputs of later layers into earlier layers. As a result, the network is able to generate ongoing signal activity alongside, or even in the absence of, input patterns. (Churchland, P.M. 1995, pp. 97–121) These nets have proved successful in modeling the recognition of sentences with quite complex grammatical structures, such as those with multiple embedded relative clauses. Feeding back activation levels evidently enables the net to keep track of shifting contexts of usage. In the domain of human behaviour, the prototypical processes identified by recurrent networks are patterns of activity extended through time. They can include sequences such as walking, blinking an eye, shaking hands,

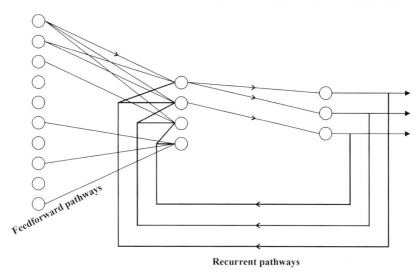

Feedforward pathways

**Recurrent pathways**

**Figure 2.2** A three-layer recurrent net. The recurrent pathways from the output layer permit these signals to be combined with new feedforward signals reaching the hidden layer. The addition of recurrent pathways gives nets a short-term memory. Not all connections are shown.

pronouncing words, driving a golf ball, and a host of socially conditioned gestures and expressions. Recurrent networks underwrite such basic temporal patterns as heartbeat and breathing. (Churchland, P.M., 1995 pp. 97–114)

The addition of a feedback loop introduces another important distinction for characterizing ANNs. Learning by backpropagation, as shown in Figure 2.1, is an example of *supervised* learning because the error signal is generated by a mismatch between the net's output and some target output that functions as a teacher. While this can occur with recurrent networks too, the existence of a feedback loop functioning as a way of presenting again the network's past experience can also allow the possibility of *unsupervised* learning, or convergent weight adjustment, to occur. By making the target output the next input, the recycling of past processing permits 'self-supervised learning' in the sense that the net is learning to predict its future inputs: '…the prediction task requires no special teacher, since the target output is simply the next input. All that is required to be psychologically plausible is to assume that this processing can lag a few steps behind the actual input' (Elman *et al.* 1997, p. 81).

Understanding knowledge representation in terms of the strength and geometry of weights in a distributed network has profound consequences for familiar ways of thinking in a number of fields. One venerable, if controversial distinction, is the alleged bifurcation between 'knowing that' and 'knowing how'. Roughly speaking, the former expresses propositional knowledge, the latter, knowledge of skills. Sentential models of knowledge representation are good on propositions but poor on skills. It is often possible to give linguistic descriptions of non-sentential expert performance, but care must be taken in drawing inferences about what a person might know. Consider the skill of speaking grammatical English. Suppose there exists a set of rules for describing all and only the grammatical sentences of English. We can say that these rules *fit* the language. Assuming knowledge of these rules is not explicit, can we say that the skill of speaking English depends on tacit knowledge of these rules? A broadly Chomskyan approach to language accepts this assumption. That is, the rules also causally *guide* the acquisition and production of English. In this way what is spoken of as a skill can be transformed into an instance of knowing that, where the relevant propositions are rules, in this case the rules of language implicitly, or tacitly, known.

Quine (1972) has raised one difficulty with this inference to tacit knowledge. He notes that we can construct extensionally equivalent grammars for a language: different sets of rules that all generate the same grammatical sentences. But while many sets may fit a language, presumably only one guides it. The methodological puzzle is to demarcate the guiding set when the linguistic evidence runs only as deep as fitting. Once we move beyond linguistic evidence to the neurophysiology of language processing, however, neural network representations of language raise the possibility that tacit knowledge of language is not rule based at all. For neural nets are not rule based. If causal guiding representation is extensionally equivalent to rule based fitting, then language simply appears *as if* it were rule based.

The point of this example is to highlight two claims. First, we want to say that all 'knowing that' is really 'knowing how'. And second, we want to defend this assertion even for paradigmatic symbolic representations of knowledge, such as is usually assumed for natural languages. Knowing that $p$, where $p$ is some linguistically expressed proposition such as 'today is payday', should be understood as knowing how to produce the quoted linguistic string, or token. This know-how will also include knowledge of the conditions under which it is appropriate to produce this physical-symbol token.

The contrast between grammar and networks in language can be made in relation to theory and practice in other skilled performances. Take such complex practices as successful teaching, or administration. Presumably, relevant theory in these domains would express in linguistic formulations significant regularities associated with these practices. Not surprisingly, the business of extracting context invariant features from highly context dependent events has proved difficult. And yet without the linguistic theory formulations at hand people can learn through experience to become good teachers or administrators. They may even develop an espoused theory to explain their actions. Nevertheless, we now have the option of seeing learning through doing in terms of steady weight adjustments necessary to minimize the gap between expectation and experience. A person's theory-in-use can be regarded as a configuration of weights which causes, or guides, expectation driven behaviour. Learning will be epistemically progressive where coherent adjustments to the net minimize the gap between feedforward expectation and feedback from experience.

A consequence of seeing knowledge of practice as non-linguistic, causally encoded configurations of weights, is that a new theory of competence is required. (Clark

1993, pp. 42–67) One familiar view imposes standards of symbolic manipulation: we have to be able to explain satisfactorily what we are doing and why we are doing it. At the other extreme is the 'no theory' approach: it is sufficient merely to demonstrate skilled performance. The first view distracts attention from learning by trading in relations among symbolic representations. The second bypasses reference to cognition altogether. In the middle can be found some combination of reflection and action with the model of the reflective practitioner perhaps the best known in educational studies. An advantage of having a sound inner workings account of competence is that it enables articulation with a theory of learning to be competent, essential for educating for professional practice.

## Prototypes, Explanation and Competence

In developing a theory of practical knowledge to the point where it can account for professional competence, we need to do more than just outline some general features of naturalistic knowledge representation. We need to say something about the representation of particular skills. We do this by way of elaborating a naturalistic view of prototypes, indicating how their physical instantiation might be modeled in ANNs. We proceed by first discussing an example – the identification of prototypical features of expert (as opposed to novice) teaching. Agreeing with Sternberg and Horvath, (1995, p. 10) that instances of teaching form a category, where each instance possesses certain features (to some degree), then a 'prototype may be thought of as the central tendency of feature values across all valid members of the category'. Similarity based prototypes are more flexible and 'fuzzy' and are better able to represent the graded nature of human performance and judgement. For many teaching practices, the distinction between expert and novice teaching is not readily formulable in terms of the more brittle 'all or nothing' resources of language. But while it might be difficult to characterize this distinction in terms of a traditional competence theory of expertise, neural networks are typically structured to enable prototype analyses of knowledge.

Consider again the feedforward net of Figure 2.1. Let us suppose that the thirteen input nodes are used to process a thirteen place input column vector where each element stands for a numerical value between 0 and 1 corresponding to the degree to which each of thirteen features of expertise nominated by Sternberg and Horvath (1995, p. 15) is present in a teaching act. The first five elements code for knowledge:

- content knowledge
- pedagogical knowledge (content specific)
- pedagogical knowledge (content non-specific)
- practical knowledge (explicit)
- practical knowledge (tacit).

The next five code for efficiency:

- automatization
- executive control (planning)

- executive control (monitoring)
- executive control (evaluating)
- reinvestment of cognitive resources.

The last three code for insight:

- selective encoding
- selective combination
- selective comparison.

We need not be concerned with the precise details of each of these features, or how raters of the suitably individuated teaching acts made their ratings. For the degree of presence or absence of these features, one can assume some Likert scale judgement, for example. We just need to suppose that, say, fifty teaching acts have been coded — twenty five performed by experts and twenty five performed by novices. Notice that the output nodes in Figure 2.1 are ultimately connected to a two place output vector.

Let us suppose that if an input vector codes for expert teaching, that is, it codes a teaching act performed by an expert teacher, then the correct output is [1,0] indicating teaching done by an expert, and for the inputting of novice teaching features we take the correct output to be [0,1]. Now feed in the fifty vectors, in random order, into the net. Since the initial setting of weights is arbitrary, the output produced for each input pattern will almost certainly be wrong. However, the correct output is also presented for each pattern so the net's error can be computed and the weights contributing to this error adjusted by backpropagation. After about twenty passes through the entire data set of fifty patterns, or vectors, it is likely that the net will begin to converge on the correct output. When fully trained, the net will settle into a set of weights that will correctly classify all fifty vector codings for teaching acts into the two categories of expert and novice. It should also be able to classify correctly additional vectors outside its training schedule. In other words, it now has a firm grasp of the distinction between expert and novice teaching acts though, of course, without the benefit of a linguistically expressed rule.

One way in which the net's representation of these concepts can be construed is in terms of the properties of the hidden layer of five nodes. After training, each input vector produces a characteristic pattern of activation across these nodes. Since all outputs from this layer partition into one of two classifications, there are just two such characteristic patterns. See Figure 2.3 below. Each can be regarded as a *prototypical activation pattern* — one for expert teaching and one for novice teaching. (Churchland 1989, pp. 200–218) Each prototype, physically shaped under the pressure to produce one of two possible correct outputs, functions as a template for classifying inputs into similarity classes. Another way to view it is as a way of *compressing* the data comprising the set of inputs. If the hidden layer contains too many nodes, the net will just learn each input-output pair separately and fail to generalize correctly in a way that will enable it to classify new data. It will just function as a look-up table. On the other hand, if there are too few nodes in the

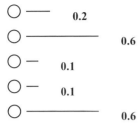

(a) Hidden layer activation levels for teaching acts performed by an expert.

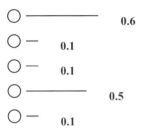

(b) Hidden layer activation levels for teaching acts performed by a novice.

**Figure 2.3** Hidden layer prototypical activation patterns for teaching acts performed by (a) expert and (b) novice teachers.

hidden layer, it will lack the resources to capture all the regularities, perhaps even failing in its learning to converge on a target acceptable error rate. Needless to say, getting this aspect of design right is not an exact science.

Just as a traditional competence theory would explain an act as expert by deriving it from the relevant linguistic formulation of what constitutes expertise, so explanation in neural network terms consists in subsuming instances under appropriate prototypes. The analysis of activation patterns on hidden layers has become an important tool in probing the understanding of trained networks. (See also Chapter 10.) Indeed, it would be an interesting exercise to train up a network on data for a cluster analysis of the hidden layer expressing the Sternberg and Horvath taxonomy of features.

What we have said for the recognition of expertise in teaching, of course applies to the *production* or *conduct* of expert teaching. In this case, the teacher's theory-in-use would be the host of activation patterns that guide the teacher in classifying the flux of classroom activity into particular similarity groupings, and then initiate appropriate clusters of teacher responses. Because these responses are actually processes, they are best theorized in terms of prototypical trajectories identified by recurrent networks.

Following Churchland (1989, pp. 197–223) we can construe explanation for all natural cognizing creatures as subsumption under a prototype:

> Explanatory understanding consists in the activation of a specific prototype vector in a well-trained network. It consists in the apprehension of the problematic case as an instance of a general type, *a type for which the creature has a detailed and well–informed representation*. Such a representation allows the creature to anticipate aspects of the case so far unperceived, and to deploy practical techniques appropriate to the case at hand. (Churchland 1989, p. 210, his emphasis.)

Since there are many different kinds of prototypes, explanations will vary. To use some of Churchland's examples, we might explain why an animal is spotted by observing that it's a leopard and all leopards are spotted. This is an example of a 'property-cluster prototype'. (Churchland 1989, pp. 212–213) Temporal sequences associated with kinds of causal prototypes are also common: e.g. 'Why did the water boil? Because it was exposed to heat'. Much functional explanation involves practical prototypes. Objects are grouped, not according to their make-up, or constituent structure, but according to their purpose. The term 'pump' is an example of this. More to the point for our purposes, there are many different 'social-interaction prototypes'. (Churchland 1989, pp. 216–217) These can capture central tendencies in learning about people's rights, the nature of confidentiality, when a person should be promoted, or fired, the grading of essay assignments, the distribution of praise, and the identification of administrative problems and characteristic suitable responses. (See especially Chapter 7 for a more extended discussion of social and moral prototypes.)

Granted this non-sentential account of explanation, we can see more clearly how to characterize practical competence in non-sentential terms. In the brief sketch of our naturalistic, coherentist, theory of knowledge given in Chapter 1 (though expressed systematically in Evers and Lakomski 1991, pp. 19–45, and 1996, pp. 29–40, 115–128, 238–246, 262–270) the justification of knowledge claims is a matter of global inference to the best explanation. That is, a claim is accepted if its inclusion into our global theory leads to that theory enjoying more of the epistemic virtues of coherence. The aim is to produce a kind of global 'best fit' for all our theoretically informed experiences. Now, if this epistemology is itself naturalized in terms of the dynamics of neural networks, then the notion of inference to the best explanation is ideally replaced by the notion of 'activation of the most appropriate prototype vector'. (Churchland 1989, p. 218) Happily, this latter notion extends over *actions* because prototype vectors exist for temporally extended processes such as complex pieces of behaviour. Competent practice therefore comes out as a matter of activating the most appropriate prototype vectorial representation of prototypical *processes*. Sequences of informed, intelligent action, presume a causal, inclusive, cognitive dynamics that results in the triggering of a suitable prototypical process vector. Despite a lack of detail, notice how the advantages of coherence tell in favour of some representationally rich causal model of practical knowledge in order to avoid the miraculous coincidence problem.

While a very large amount of practical and professional knowledge is best construed as a matter of pattern processing rather than the logical manipulation of sentential structures, we are not denying the value of the sentential. Clearly, if we have useful language-like formulations of experience at hand these can be of tremendous value especially if we are able to manipulate the formulations in ways that enable them to go proxy for experience, for example, in the exploration of counterfactuals. A major point we want to emphasize, however, is that in order to link these symbolic representations up with practice, there is much merit in reconstruing them as patterns of input to a physically realistic model of the human brain. It is just not feasible to develop high level theories of practice, and the representation of practical knowledge, that do not articulate smoothly with the substrate of lower level causal activity required to sustain embodied action.

The ideas discussed in this chapter concern cognitive processes in individuals. However, much cognitive activity in administrative and organizational contexts needs to be seen as a collective phenomenon. Augmenting our naturalistic account of practice in this direction, we therefore give some attention, in the next chapter, to the socially distributed nature of cognition, with a special focus on organizational learning in Chapter 5. This follows an account of the socially relational nature of leadership practices in Chapter 4. Some theoretical extensions of this chapter's analysis of practical knowledge is advanced in the concluding chapters on research, where we construe epistemology in general as a model for research methodology, and our naturalism in particular as a tool for the identification and characterization of social and cultural patterns in organizational life.

## References

Aristotle. (1962). *Nicomachean Ethics*. (Indianapolis: Bobbs-Merrill).

Bechtel W. and Abrahamsen A. (1991). *Connectionism and the Mind*. (Oxford: Basil Blackwell).

Caudill M. and Butler C. (1992). *Understanding Neural Networks*, Volumes 1 and 2. (Cambridge, MA: MIT Press).

Churchland P.M. (1989). *A Neurocomputational Perspective*. (Cambridge, MA: MIT Press).

Churchland P.M. (1995). *The Engine of Reason, the Seat of the Soul*. (Cambridge, MA: MIT Press).

Churchland P.S. (1995). Can neurobiology teach us anything about consciousness?, in Morowitz H. and Singer J.L. (eds.) *The Mind, The Brain, and Complex Adaptive Systems*. (New York: Addison-Wesley). pp. 25–43.

Clark A. (1993). *Associative Engines*. (Cambridge, MA: MIT Press).

Cussins A. (1992). The limitations of pluralism, in Charles D. and Lennon K. (eds.) *Reduction Explanation and Realism*. (Oxford: Clarendon Press).

Dewey J. (1916). *Democracy and Education*. (New York: Free Press).

Dewey J. (1938). *Experience and Education*. (New York: Macmillan).

Elman J.L. *et al.* (1997). *Rethinking Innateness: A Connectionist Perspective on Development*. (Cambridge, MA: MIT Press).

Evers C.W. and Lakomski G. (1991). *Knowing Educational Administration*. (Oxford: Pergamon Press).

Evers C.W. and Lakomski G. (1996). *Exploring Educational Administration*. (Oxford: Pergamon Press).

Freire P. (1972). *Pedagogy of the Oppressed*. (London: Panguin Books).

Hoy W.K. and Miskel C.G. (1996). *Educational Administration: Theory, Research and Practice*. (New York: McGraw-Hill, Inc.).

McClelland J.L. and Rumelhart D.E. (1988). *Explorations in Parallel Distributed Processing*. (Cambridge, MA: MIT Press).

Plunkett K. and Elman J.L. (1997). *Exercises in Rethinking Innateness: A Handbook for Connectionist Simulations*. (Cambridge, MA: MIT Press).

Popper K.R. (1959). *The Logic of Scientific Discovery*. (London: Hutchinson).

Popper K.R. (1963). Science: conjectures and refutations, in K. R. Popper *Conjectures and Refutations: the Growth of Scientific Knowledge*. (London: Routledge and Kegan Paul).

Popper K.R. (1969). Epistemology without a knowing subject, in J.H. Gill (ed.) *Philosophy Today*, Number 2. (Toronto: Macmillan).

Quine W.V. (1972). Methodological reflections on current linguistic theory, in Davidson D. and Harman G. (eds.) *Semantics and Natural Language*. (Dordrecht: D. Reidel).

Rumelhart D.E. and McClelland J.L. (eds.) (1986). *Parallel Distributed Processing*, Volumes 1 and 2. (Cambridge, MA: MIT Press).

Schön D. (1983). *The Reflective Practitioner: How Professionals Think in Action*. (New York: Basic Books).

Schön D. (1987a). *Educating the Reflective Practitioner: Toward a New Design for Teaching and Learning in the Professions*. (San Franciso: Jossey-Bass).

Schön D. (1987b). Educating the reflective practitioner, Invited Address, American Educational Research Association, Washington, DC, http://educ.queensu.ca/~ar/schon87.htm.

Simon H.A. (1976). *Administrative Behavior*. (New York: The Free Press, 3rd Edition).

Simon H.A. (1995). New decomposability and complexity: how a mind resides in a brain, in Morowitz H. and Singer J.L. (eds.) *The Mind, The Brain, and Complex Adaptive Systems*. (New York: Addison-Wesley). pp. 25–43.

Sternberg R.J. and Horvath, J.A. (1995). A prototype view of expert teaching, *Educational Researcher*, 24(6), pp. 9–17.

Ward Systems Group (1993). *NeuroShell 2*. (Frederick M.D.: Ward Systems Group, Inc).

# 3

# Socially Distributed Cognition

In the preceding two chapters we laid out some of the general framework for understanding the representation of practical knowledge. With this chapter, we conclude Part I of our book which maps out our views on Cognition and Practice. We here continue our discussion of naturalistic practice — whether in education or educational administration — and concentrate on the important fact that practice is always *social* practice. It is embedded in cultural, situation specific contexts, engaged in by agents in interaction with other agents and their environments, both social and natural. We offer an expanded view of human cognition which unlike traditional cognitive science models, rejects the idea that the sum total of human cognition and intelligence resides in the ability to process symbols in our individual heads. We are aided in this enterprise by contemporary connectionists who have begun to explore cognition beyond the individual skull, and to consider cognition as *distributed* between other knowers and their material contexts. Hence, the interactive and reciprocal relationships between cognition and culture have become a prominent and new avenue for cognitive science research.

It is important to acknowledge here that the individual ownership model of cognition has also come under attack from a quite different but intersecting field. *Situated action* is an umbrella description for a number of approaches, and derives mainly from cultural anthropology, ethnomethodology, discourse analysis, and interpretive social science generally. Given its social anthropological and cultural roots, *situated action* stresses the importance of the construction of human cognition in the everyday cognitive practices of humans and rejects the view of cognition as uniquely symbol processing.

The most important contribution of this perspective is its insistence on the cultural and contextual features as integral parts of cognition, demonstrated in various empirical studies including the learning of school mathematics. This perspective has recently become prominent in the educational research literature and can be considered the most prominent challenge, coming from education, to the symbol processing or cognitivist view. However, in so far as *situated action* remains largely agnostic about the fine-grained detail of the causal mechanisms of cognition, and in so far as connectionism does combine a causal account with *situated action*'s cultural perspective *and* includes symbol processing, it extends human knowledge of cognition into areas yet unexplored. It promises to provide a far more comprehensive framework for the explanation of human thought and action, in education as elsewhere. In our view, since *situated action* correctly stresses the cultural-social aspects of cognition, and is by far the most sophisticated approach in the education literature to attempt to explain human thought and practice, a closer examination of its claims is well warranted.

The following sections discuss the main claims and features of *situated action* over against the classical definition of symbol processing, Newell and Simon's physical symbol system hypothesis to which we referred in Chapter 1. In particular, since the proponents of either perspective engage one another directly in two separate debates, in the cognitive science and education literature respectively, a brief account of these is given to focus on their differences as well as agreements. The detailed nature of these differences provides a more comprehensive picture of the scope of what we might call *cultural connectionism*, and indicate where we believe research efforts ought to be expanded in future.

## Cognition as Situated Action

*Situated action* is a perspective which covers a range of views and terminologies under headings such as 'situated activity'; 'situated social practice'; 'situated learning'; 'situated' or 'distributed cognition'. The collections of essays in Rogoff and Lave 1984, and Resnick, Levine and Teasley 1991 provide good examples of these different views. It is not possible here to do justice to the complexity and diversity of *situated action* approaches which range from social psychological, anthropological, ethnomethodological and discourse analysis perspectives to those embedded in cognitive psychology and cognitive science. Attempts to identify claims as central to such a perspective need to be mindful of the theoretical contexts in which the claims figure.

However, there is at least one point of agreement in this diversity in virtue of which *situated action* qualifies as a distinct position. It is opposed to the central claim of the physical symbol system hypothesis that symbol processing is at the core of human intelligence and cognition. While most *situated action* theorists do not deny that symbol processing has a role to play in human cognitive activity, different positions are taken about what precisely this role is presumed to be. The importance and function of symbol processing in human cognition is an extraordinarily difficult and complex issue which is as yet not well understood. In the present context, some very recent developments in the artificial neural net connectionist literature are sketched to provide a glimpse of work in progress.

To begin the discussion of *situated action*, Lave's description of situated social practice, or where appropriate situated learning, provides a good starting point since she is one of *situated action*'s most frequently cited and discussed theorists (Rogoff and Lave 1984; Lave 1988, 1991; Lave and Wenger 1991; Chaiklin and Lave 1993).

Her broad description is characteristic of the social-cultural flavor of *situated action*. Describing her project as a social anthropology of cognition, Lave is concerned to emphasize the fluid boundary between intra- and extra-cranial human experience, a boundary more characterized by 'reciprocal, recursive, and transformational partial incorporations of person and world in each other within a complex field of relations between them.' (Lave 1988, p. 1). 'The point' she says,

> ...is not so much that arrangements of knowledge in the head correspond in a complicated way to the social world outside the head, but that they are socially organized in such a fashion as to be indivisible. 'Cognition' observed in everyday practice is distributed — stretched over, not divided among — mind, body, activity and culturally organized settings (which include other actors).
> (See also Chaiklin and Lave 1993, p. 17)

This view not only maintains that cognition is distributed, but that the correct focal point for the study of human cognition is 'out there' in the cognitive performance of people's everyday routines. Such a perspective is opposed to studying cognition in laboratory or other experimental settings which specify and test only certain kinds of specialized cognitive tasks. One of her early examples are empirical studies of arithmetic as cognitive practice, the Adult Math Project (AMP), conducted in such everyday contexts as supermarket shopping, and beginning a dieting program, with the goal of establishing how well formally taught school arithmetic transfers to practical everyday contexts. Lave finds that mathematical skills taught in the school context appear not to be used in everyday reasoning contexts.

According to Lave's position, acquiring knowledge and skill, and becoming a master practitioner, is a matter of ongoing practice in communities of practice, exemplified in, for example, some forms of apprenticeships which do not rely on formal teaching. The danger of the traditionally held cognitivist view is that it dis-embeds knowledge/cognition from the routine contexts of its use: 'extraction of knowledge from the particulars of experience, of activity from its context, is the condition for making knowledge available for general application in all situations.' (Lave 1988, p. 8). In her view formal schooling epitomizes this disembedding. Hence, she concludes, learning in schools 'is most likely to fail'. (Lave 1991, p. 80) The continuity of learning over different settings is, in cognitive psychology, attributable to the explanatory concept of transfer of learning whose cogitivist definition Lave rejects. *Situated action*, on her account, provides a better explanation of the internalization of knowledge, and of transfer than that offered by cognitive psychology.

While Lave's approach is more colored by critical social ethnography and anthropology, writers such as Greeno and Moore (1993, 1997), and Clancey (1993) can perhaps be located more directly in the cognitive science tradition. For Greeno and Moore (1993, pp. 49–50), for example, *situated action* or situativity theory claims

> ...that cognitive activities should be understood primarily as interactions between agents and physical systems and with other people. Symbols are often important

parts of the situation the people interact with, and understanding how symbols are used and constructed in activities is one of the major problems of cognitive theory.

Greeno (1997, p. 15) believes that *situated action* is more useful in building a more comprehensive theory of human cognition than the physical symbol system hypothesis in that it 'considers processes of interaction as basic and explains individual cognitions and other behaviors in terms of their contributions to interactive systems.' Methodologically this implies studying the activities of groups to detect the properties of these social practices which provide the main explanatory concepts of *situated action*. (For examples see Hastie and Pennington 1991; Scribner 1984; Levine and Moreland 1991; Hutchins 1991.) A useful summary of the main claims of *situated action* is offered by Clancey (1993, pp. 90–93) who, unlike his social anthropology colleagues attempts to give a comprehensive rather than oppositional picture of both symbol processing and *situated action* in neuropsychological terms. *Situated action* research, Clancey (1993, p. 92) states, is '... Not merely about an agent 'located in the environment' 'strongly interactive', or 'real time'... rather [it is] a claim about the internal mechanism that coordinates sensory and motor systems'. His formulation makes it clear that he wishes to integrate symbol processing into a kind of neuropsychological version of *situated action*.

In introducing the debate between the two sides in the 1993 Special Issue of *Cognitive Science,* the editor formulated the core issue like this: 'Is symbolic cognition a special case of cognitive activity (the position Greeno and Moore hold), or is *situated action* a special case of symbolic cognitive ability (the position that Vera and Simon hold)?' (Norman 1993, p. 5). Drawing on both the 1993 *Cognitive Science* debate and the subsequent publication of another critical exchange between Simon and colleagues and Greeno and colleagues in the 1996 and 1997 issues of *Educational Researcher,* the nature of the disagreements between these two approaches come to light even more clearly; for a parallel critical discussion regarding teaching see Packer and Winne 1994. For Greeno (1997, p. 5), the issue is fundamentally about which 'framework offers the better prospect for developing a unified scientific account of activity considered from both social and individual points of view, and which framework supports research that will inform discussions of educational practice more productively.'

Of particular interest in the present discussion are the contributions by Vera and Simon (1993a) and Anderson, Reder and Simon (1996) which, as critics of *situated action*, in turn provide the foil for critique by Greeno and Moore (1993), Greeno (1997) and Clancey (1993). Given the above statements, the positions advocated and defended are not considered amenable to compromise with the possible exception of Clancey. Vera and Simon (1993a) argue that advocates of *situated action* have nothing new to contribute. Thus, they do not threaten the symbol processing view which is able to contain the claims and arguments of the opposition. Greeno *et al.*, for their part, contest their argument believing their approach to be superior in that they can accommodate symbol processing in their broader account. Before it is possible to consider what a preliminary resolution of this issue might look like, it is necessary to understand better the nature and issues of disagreement. In order to do so it is helpful to take a closer look at what is meant by the physical symbol system hypothesis which is at the heart of classical artificial intelligence and which until very recently has defined how to understand human cognition, reason, and intelligence.

## Cognition as Symbol Processing

Put in its simplest form, the physical symbol system hypothesis maintains that human knowledge and rationality consist in the ability to manipulate (linguistic and quasi-linguistic) symbols in the head. The emphasis is on the internal processing structures of the brain and the symbolic representations of mind. Discussed and defended extensively (Newell and Simon 1972; Newell and Simon 1976; Simon 1979; Anderson 1983), it is worthwhile to quote Simon (1990, p. 3) in full. The physical symbol system hypothesis

> states that a system will be capable of intelligent behavior if and only if it is a physical symbol system.... [it] is a system capable of inputting, outputting, storing, and modifying symbol structures, and of carrying out some of these actions in response to the symbols themselves. 'Symbols' are any kinds of patterns on which these operations can be performed, where some of the patterns denote actions (that is, serve as commands or instructions).... Information processing psychology claims that intelligence is achievable by physical symbol systems and only such systems. From that claim follow two empirically testable hypotheses: 1. that computers can be programmed to think, and 2. that the human brain is (at least) a physical symbol system. (See also Vera and Simon 1993a, pp. 8–11; Vera and Simon 1993b).

In order to test these hypotheses, a computer is programmed to perform the same tasks used to judge how well humans think, and to demonstrate that the computer uses the same processes humans employ in performing the tasks. Simon and his colleagues use 'thinking-aloud protocols, records of eye movements, reaction times, and many other kinds of data as evidence' (Simon 1990, p. 3).

In computers, symbols are typically patterns of electromagnetism but their physical nature differs depending on the kind of computer. In any case, Simon does not believe that their physical nature matters: what matters is their role in behavior. He observes that humans do not actually know how symbols are represented in the brain other than that they are likely to be patterns of some kinds of neuronal arrangements (Vera and Simon 1993a, p. 9). Another important feature of patterns as symbols is their ability to designate or denote, either other symbols, patterns of sensory stimuli or motor actions: 'Perceptual and motor processes connect the symbol system with its environment, providing it with its semantics, the operational definitions of its symbols' (Vera and Simon 1993a, p. 9). While the social environment is considered to some extent, the explanation of how and in what form new knowledge is stored in memory, how learning occurs, can be studied 'primarily in terms of its internal mechanisms ... taking the input (e.g., material from a textbook) as given, and seeking to model how that input changes the internal contents of memory so that the system will subsequently possess the desired skill or the desired knowledge' (Vera and Simon 1993a, p. 43). Simon insists that symbols, and hence symbolic processing, must not be identified as exclusively verbal or linguistic processing. Rather symbols may 'designate words, mental pictures, or diagrams, as well as other representations of information' (Vera and Simon 1993a, p. 10).

Although the physical symbol system hypothesis has been highly successful in that it 'has been tested so extensively over the past 30 years that it can now be regarded as fully established, although over less than the whole gamut of activities that are called 'thinking'' (Simon 1990, p. 3), Simon concedes a number of unresolved issues. One of the most important is how the brain's symbol processing capacities are realized physiologically. He notes that while information processing psychology says much about the 'software of thinking', it has little to say about its 'hardware' or 'wetware', a gap which needs to be filled (Simon 1990, p. 5; see also Newell, Rosenbloom, and Laird 1996).

It is also acknowledged that there are computational limits in information processing in terms of both speed and organization of a system's computations and sizes of its memories. The example of playing a perfect game of chess is instructive here. Such a feat would require examining more chess positions than there are molecules in the universe. Simon draws the following conclusion, 'If the game of chess, limited to its 64 squares and 6 kind of pieces, is beyond exact computation, then we may expect the same of almost any real-world problem, including almost any problem of everyday life.' (Simon 1990, p. 6). On the basis of this observation, he concludes that '... *intelligent systems must use approximate methods to handle most tasks. Their rationality is bounded*' (Simon 1990, p. 6; emphases in original). The doctrine of bounded rationality is arguably the most influential feature of mainstream organization and (educational) administration theory, which, as a theory of rationality, has determined all the major features of organizational functioning.

Finally, in their assessment of the success of the perspective that they have been instrumental in shaping, Newell, Rosenbloom and Laird (1996, p. 127) single out some items as yet unknown regarding their effects on cognitive architecture. These are the issues (1) of 'acquiring capabilities through development, of living autonomously in a social community, and of exhibiting self-awareness and a sense of self ... '; (2) how to square their cognitive architecture with biological evolution which puts a premium on perceptual and motor systems; and (3) how to integrate emotion, feeling and affect into cognitive architecture. These comments show clearly the boundary drawn between cognitive architecture and the world although the exact drawing of it is accepted as problematic.

## Situated Action versus Symbol Processing

The difficulty of presenting the arguments of both perspectives is that either side believes that its view can accommodate the perspective of the opponent when suitably redescribed. Vera and Simon (1993a, p. 46) conclude their account of *situated action* 'without finding reasons why such action cannot be accommodated within the physical symbol-system hypothesis' and that 'SA [situated action] is not a new approach to cognition, much less a new school of cognitive psychology'. Greeno and Moore (1993, p. 55) in turn deny the centrality of the physical symbol system hypothesis for human intelligence, but agree that 'cognition includes symbolic processing, but they are not coextensive'. In addition, each side accuses the other of misrepresenting at least some aspects of their position, resulting in continual adjustments, redefinitions, and misunderstandings. It is clear that their differing theoretical backgrounds and philosophical assumptions account largely for the language in which the arguments are couched. An attempt is made in the following pages to present both sides to demonstrate where they differ from each other and also point to possible agreements.

In their critical review of *situated action*, Anderson *et al.* (1996, p. 5) assert that the following four claims are central:

1.  Action is grounded in the concrete situation in which it occurs;
2.  Knowledge does not transfer between tasks;
3.  Training by abstraction is of little use; and
4.  Instruction must be done in complex, social environments.

The authors agree with the weak version of claim (1) in that much of what is learnt is specific to the situation but they reject the strong version that all knowledge is situation-specific. The latter version gives rise to the claim (2) that more general knowledge cannot transfer to other contexts, such as arithmetic formally learnt in school contexts. They take Lave's (1988) position to be an example of the strong version. However, whether one assumes the weak or strong versions of lack of transfer between contexts, there is a large amount of empirical evidence in the psychological literature which shows that degrees of transfer — from large to modest to none to negative — is a function of 'representation and degree of practice; of the experimental situation and the relation of the material originally learned to the transfer material', and on 'where attention is directed during learning or at transfer' (Anderson *et al.* 1996, pp. 7–8).

A corollary of claim (2), claim (3), if interpreted in its strong version, undercuts the legitimacy of school learning in that it leads to a demand for apprenticeship training in real-world contexts on the assumption that what is learnt in school contexts is too abstract for the job at hand. A weak version could consist in training by modeling, a more traditional pedagogical device. Anderson *et al.* (1996) provide further empirical studies which qualify the claim that abstract instruction is of little help. The literature of modern information processing appears to suggest that a combination of abstract instruction and concrete task execution is best for learning to take place. As for claim (4) that learning is inherently a social activity and should be done in complex social situations, the authors point out that while social aspects of jobs are important, skills can well be taught independently of the social context. Part training is often more effective where the part is practically independent of the larger task. This does not deny that it is useful and necessary in some situations to practice skills in their complex contexts, such as playing the violin in an orchestra. Anderson *et al.* (1996, p. 10) conclude their critical examination of the four claims attributed to *situated action* by noting that

> [I]n general, situated learning focuses on some well-documented phenomena in cognitive psychology and ignores many others: while cognition is partly context-dependent, it is also partly context-independent; while there are dramatic failures of transfer, there are also dramatic successes; while concrete instruction helps, abstract instruction also helps; while some performances benefit from training in a social context, others do not.

Greeno's (1997) reply exemplifies the different discourses from within which learning and cognition are addressed. He notes that the differences between the two perspectives are really differences between explanatory frameworks. He does not accept the claims Anderson and his colleagues describe as characteristic of *situated action*, and he attributes their formulation to the different framing assumptions of the cognitivist perspective. While the latter begins on the assumption of a theory of individual cognition, supplemented by further analyses of additional components which serve as contexts, *situated action* begins from the assumption of a theory of social and ecological interaction which leads it towards a more comprehensive theory of analyses of information structures in the contents of people's interactions. These basic assumptions clearly show in the answers Greeno gives to the criticisms raised.

*Situated action* starts with interactive systems, including people and other material and representational systems. Where the physical symbol system hypothesis advocates the acquisition of knowledge in form of abstract representations and procedures for application in many contexts, *situated action* emphasizes learning to participate in interactions in ways that succeed over a broad range of situations. The implications for theory are that the physical symbol system hypothesis is wrongly committed to a 'factoring assumption' in that properties of individual mental processes can be analyzed and that the properties of other systems can be taken as contexts in which these processes and structures function (Greeno 1997, p. 7). As such, Greeno also admits that the cognitive perspective has been very successful. It is however 'unacceptably incomplete' in that it is not possible in the cognitive view to specify the contribution other systems have made in an individual's construction of knowledge in interaction with the environment. For *situated action*, more effective participation in groups across a wide range of situations is the relevant criterion for learning having taken place. This is juxtaposed to the formal testing of knowledge retention, attributed to the cognitive view.

Replying to Anderson *et al.*'s claim (4), Greeno comments that the issue is not one of the acquisition of skills but of the appropriate arrangement of the social conditions of learning. Learning always happens in complex social environments, such as the social arrangements which produced both textbook and computer. The educational implications of either position are that for the cognitive perspective it becomes central how to arrange for the collection of skills and understandings to be acquired most efficiently. For *situated action* alternative arrangements for learning need to consider 'the values of having students learn to participate in the practices of learning that those arrangements afford' (Greeno 1997, p. 10). This might mean beginning with relatively more complicated arrangements since learning sequences are considered 'trajectories of participation' rather than sequences of knowledge and skill acquisition of increasing complexity.

Regarding claim (2) in Anderson and his colleagues' critique, Greeno rejects the assertion that transfer of learning is antithetical to *situated action*. Rather, the issue is one of appropriately framing the question of generality and transfer, not one of discussing whether or not transfer occurs. In Greeno's conception, a test for transfer 'involves transforming the situation in which an activity was learned. To succeed in the transfer test, the activity — that is, the interaction of the learner with the other systems in the situation — has to be transformed in a way that depends on how the situation is transformed'. (Greeno 1997, p. 12) The question of generality of learning is very important because it relates directly to the school curriculum which is designed on the assumption that school-based knowledge is inherently general because abstract. Here Greeno refers to Lave's empirical work on school-learnt algorithms which were not used in reasoning away from school. However, he warns that the conclusion drawn by cognitive writers overshoots the mark when they allege that *situated action* claims that school-based learning cannot be used in non-school settings. What *situated action* does maintain is that a school learning situation has its own practices and characteristics of performance which differ from those of other non-school settings. Hence, he points out 'We cannot safely assume that, by learning the procedures of symbol manipulation which they have traditionally been taught in school mathematics, students' participation in other kinds of interactions will be strongly influenced'. (Greeno 1997, p. 12)

In discussing the issue of abstraction, claim (3) of Anderson and his colleagues' critique, Greeno notes that as far as *situated action* is concerned, the concept of abstraction is part of that of representation. Here, as before, it is obvious that fundamental theoretical formulations, rather than empirical data, are the main bone of contention. If considered as part of representation, then it involves 'the portability of symbolic or iconic representations that can be interpreted apart from their referents'. (Greeno 1997, p. 13) For students to learn these properly, for example, the representational systems of mathematics and science, they also need to learn the standard conventions used to interpret them. Such representational systems have a role to play in *situated action* as an aspect of social practice provided their meanings are understood. Only then do they contribute to learning as understood in *situated action*, and is sharply differentiated from learning a set of mechanical rules. Greeno

(1997, p. 13) observes that more needs to be known about the role of abstract representation in activity.

In their final word on the present exchange, Anderson and his colleagues confirm that there are some agreements between the perspectives, but continue to insist that *situated action* does not possess the 'right theoretical or experimental tools for understanding social cognition. Such understanding can only be achieved through serious attention to what goes on in the human mind, and not simply through external observation of social interaction'. (Anderson *et al.* 1997, p. 20) The differences between the two approaches are sharply in focus in the cognitive perspective's appraisal of the social which it attempts to understand 'through its residence in the mind of the individual'. The strength of the information-processing approach 'comes not just from its decision to focus on the individual but from its decision to analyze the knowledge possessed by the individual and how it is, and can be, acquired'. (Anderson *et al.* 1997, p. 21)

The two perspectives, as has been amply documented, come from different disciplines and research traditions, and begin from diverse starting points. Yet despite such ostensive differences and outright disagreements both approaches deliver valuable insights into human cognition and its function in the world. Weaving these into a coherent account of cognition, however, requires a closer look at symbol processing in light of the challenge posed by the connectionist artificial neural networks account. Ironically, it is through work done in this branch of connectionist neuroscience that it is possible to foreshadow a productive integration of both symbol processing and the demands of *situated action* that cognition be considered a cultural phenomenon, created through reciprocal interaction with the environment, including symbol systems.

## Symbols in the Head?

The cognitive perspective, based on symbol processing as the central characteristic of human cognition and intelligence, is a rule-based approach to cognition. Simply put, the classical cognitive perspective assumes that a computer is able to simulate human cognition and information processing when it is programmed with human like rules. For a machine to be able to learn, according to this perspective, it can be fed rules for modifying or adding rules. It was not believed that the architecture of the physical brain was of fundamental significance since the higher cognitive processes of importance for the physical symbol system hypothesis — information processing, problem-solving, and planning — are at some remove from and at a higher level of abstraction than basic neuronal processes. An excellent source book of the foundations of cognitive science is Posner 1996; see also Lycan 1990 for various philosophical implications of cognitive science and connectionism.

Despite considerable success, there are remaining difficulties which appear resistant to resolution within the symbolic perspective, as pointed out earlier. What in Simon's (1990) terminology is the problem of explaining 'rationality without optimization', i.e. how such everyday accomplishments as devising university or factory schedules, for example, are possible since they go beyond the practical bounds of computation by even the fastest super computer, is believed to be more than a temporary glitch in the program. Rather, it is indicative of what Tienson (1990) calls the 'Kuhnian crisis' of GOFAI ('good old-fashioned artificial intelligence', as Haugeland 1985, phrased it). This crisis is underscored by a range of cognitive phenomena which are either inadequately explained or ignored by the physical symbol system hypothesis. Examples identified by Clancey (1993, pp. 97–98) include the following situations:

- Regularities develop in human behavior without requiring awareness of the patterns, that is, without first person representations…
- People speak idiomatically, in ways grammars indicate would be nonsensical
- We experience interest, a sense of similarity, and value before we create representations to rationalize what we see and feel…
- Know-how is at first inarticulate and disrupted by reflection…
- Immediate behavior is adapted, not merely selected from prepared possibilities…

In addition, it seems difficult for human beings to carry out chains of reasoning in their heads or to number-crunch, feats easily accomplished by the computer. On the other hand, people are extremely good at recognizing and completing patterns and retrieving memory, capabilities which pose enormous problems for rule-based systems. Since such capabilities appear to be more in keeping with evolutionary and biological developments, it is more plausible to believe that since nature solves complex tasks by drawing on overlapping systems ('swarming', see Bereiter 1991) that the brain would do likewise. Distributed cognition appears to be the better bet to cope with complex environments and tasks than the neat and orderly process adopted by the physical symbol system hypothesis.

A further worry relates to the top-down nature of rule-based systems which seem to be incompatible with either learning or evolution. As Bereiter (1991, p. 11) notes, 'If a system is already operating according to a given set of rules, there does not seem to be any way that those rules could generate a higher level rule that controlled them.' Neither evolution nor learning appear to have a place in this conception. These observations have been complemented by a growing body of psychological research (e.g. Johnson-Laird 1983). The importance of this combined research serves to indicate that human cognition is not much like the account given by the physical symbol system hypothesis since it does not resemble our natural intelligence which, after all, it was supposed to simulate. Cognition seems much messier and more haphazard than believed by the advocates of the physical symbol system hypothesis.

The additional difficulties, expressed as those of real time and graceful degradation add to the physical symbol system hypothesis's problems, as well as two further, and more potentially threatening, sets of concerns. The real time issue refers to the fact that most artificial intelligence programs require many more than 100 steps to carry out a basic cognitive task in contrast to human problem solving. Although neurons are slow, operating in times measured in milliseconds and being thus much slower than computers, they accomplish their task within about 100 steps (the '100–step rule', see Bechtel 1990, p. 259). Similarly, graceful degradation refers to the fact that the performance of natural systems runs down gradually. Persons who suffer a stroke, for example, lose some but not all mental functions, the nervous system being 'relatively fault tolerant'. A von Neumann machine, on the other hand, which is the material basis for the physical symbol system hypothesis, 'is rigid and fault intolerant, and a breakdown of one tiny component disrupts the machine's performance' (Churchland and Sejnowski 1990, p. 233).

But the more severe problem for the physical symbol system hypothesis is that it has not been possible to construct the kinds of rules that it requires, a difficulty already hinted at in Simon's chess example above. This difficulty arises in relation to two clusters of problems: (1) multiple simultaneous soft constraints, and (2) any piece of commonsense knowledge might turn out to be relevant to any task or any other piece of knowledge; the latter cluster includes the frame and the folding problem (Tienson 1990, pp. 383 *et seq.*).

The first problem is nicely illustrated by Shulman in his description of an exceptional teacher, Nancy, who

> was like a symphony conductor, posing questions, probing for alternative views, drawing out the shy while tempering the boisterous. Not much happened in the classroom that did not pass through Nancy, whose pacing and ordering, structuring and expanding, controlled the rhythm of classroom life... [Her] pattern of instruction, her style of teaching, is not uniform or predictable in some simple sense. She flexibly responds to the difficulty and character of the subject matter, the capacities of the students... and her educational purposes. She can not only conduct her orchestra from the podium, she can sit back and watch it play with virtuosity by itself. (Shulman 1987, pp. 2–3)

It is computationally impossible to write rules for a computer which could model such skilled performance since an enormous number of factors come into play which are unpredictable. And yet good practitioners/teachers seem to be able to combine such a vast range which produces what can be recognized as exceptional practice. Constraints are soft in the sense that any constraint can be violated, needing immediate responses and adjustments, without thus jeopardizing overall performance.

The second cluster has to do with commonsense understanding. In dealing with a cognitive task, it is necessary to judge what information is relevant and to find relevant information in all the knowledge already possessed. In other words, it is necessary for an individual to be able to fold knowledge together which has been acquired under different circumstances and while solving unrelated problems. Humans manage to mix-and-match effortlessly, but computers are far less capable of such feats. At the other end of the spectrum, according to Tienson (1990, p. 385) is the frame problem which is the problem of determining how much of the new information would either change, or leave unchanged, an existing set of beliefs. This implies that it is first necessary to determine which old beliefs are in any way relevant to the new information. These kinds of inferences are drawn naturally, but a computer would need to search every piece of old information to assess relevance, which is highly inefficient. The magnitude of this problem for a computer is such that even if it did have access to all of a human being's knowledge, it would need to specify rules in advance for what each piece of new information would or would not change in any given situation.

Humans never know which bit of information might be useful for the solving of what problem, but they are very good at retrieving or finding the relevant information when they need it, and also note failures. In order to do this it is necessary first to see information as relevant and then retrieve it and do so all at once. Human memory appears to be based on content, or is content addressable. This poses a problem for the physical symbol system hypothesis since in computer architecture, information is retrieved from where it is stored, from its address.

What the above discussion shows is that humans are routinely capable of cognitive feats which appear to surpass the capabilities of von Neuman machines to simulate them. Although advocates of the physical symbol system hypothesis have made some progress and developed more sophisticated programs which manage to overcome some of the biological constraints of serial processing computers were not able to overcome (e.g. Anderson's ACT and Newell's SOAR), whether such fine-tuning of the physical symbol system hypothesis is more helpful in understanding human cognition, does in the end depend on the architecture of the physical brain. And insights into its workings cannot ultimately be established by behavioral data alone.

In so far as artificial neural nets attempt to model real brain functioning characterized by massive interconnectivity and parallel distributed processing, and in so far as nets can program themselves without the benefit of rules (Rumelhart and McClelland 1986; Churchland and Sejnowski 1994), artificial neural nets can be said to provide a more productive framework for understanding human cognition. The 'ancient paradox' between those who place 'the essence of intelligent behavior in the hardness of mental competence ... ' and those who place 'it in the subtle softness of human performance' (Smolensky 1988, p. 199) can indeed be solved.

Information and knowledge in connectionist systems is actively represented in the weights between the nodes which make up a pattern of activation. Learning in such a system consists of having the weights changed. (For more detail regarding neural net architecture and its implications for education see Evers and Lakomski 1996, Chapter 9.) A representation may be brought about by the activation of a sufficient number of its nodes, but not necessarily all of them. This means that the same representation may be brought about by the activity of different nodes on different occasions (Tienson 1990, p. 391). It is also a feature of connectionist systems that weights are set so that representations can complete themselves when only a few of their nodes are active. Patterns of activation are not stored in the manner of data structures. When the information is not actively in use there is no pattern in the system; the role of symbol in a connectionist system is played by a pattern of activation. Hence, connectionist systems can also be described as the sub symbolic (Smolensky 1988). Pattern recognition, carried out with ease by neural nets, is a central cognitive function and more fundamental than the rule-based processing presumed to be the pinnacle of cognitive work.

Other important features of artificial neural nets are that they do not contain a central processor which determines how the system functions as a whole since connections between nodes are local. There is no one place in the system which knows what the system is doing as a whole, and what goes on in one part is independent from other parts. A neural net nevertheless does represent content across the system when it is in a particular state, and it can be said to have stored knowledge in the connection weights between its nodes. Networks are thus capable of internal representations, only such representations are not symbolic, they are certain patterns of activation. Hence, the cornerstone of connectionism can be expressed as 'The intuitive processor is a sub conceptual connectionist dynamical system that does not admit a complete, formal, and precise conceptual-level description' (Smolensky 1988, p. 160).

But this does not imply that connectionists eschew symbolic representation, although its significance is still being debated. Formalized knowledge in linguistic structures as the most prominent example of symbol processing indeed serves important functions: (1) it is publicly accessible; (2) different people can check its validity thereby attesting to its reliability or otherwise; and (3) its formal character, logical rules of inference, means that it is universally applicable and that people need not necessarily have to have experience in the actual domain to which it applies (Smolensky 1988, p. 153). Knowledge formalization at the cultural level, however, differs significantly from that of the individual in that it is neither publicly available nor completely reliable; it is also dependent on ample experience. In the section to follow consideration is given to how embodied and environmentally embedded cognition attempts to solve the problem.

## Embodied and Embedded Cognition

While modern connectionism is concerned to model in ever finer detail how the physical brain works, there is also another direction in which the new connectionism has been developing. This research is directed outwards to the cultural contexts in which humans exist and in which they enact their cognitive activity. In order to get a better understanding of the importance of this shift, it is first necessary to ask what led symbols to be placed in the head at all? What kinds of assumptions have made it possible for the physical symbol system hypothesis to edit culture out in its attempt to define human cognition?

The task of this section is to explore how the insights of *situated action* and symbol processing might be rendered coherent in the light of recent developments in connectionism. This attempt at synthesis provides a first step at simultaneously naturalizing culture and context and enculturating cognition (see the work of

Hutchins 1991, 1996 and Clark 1997). Such a move implies a conception of culture quite different from the traditional in that it is not considered as a collection of artifacts external to the mind, but as a process.

The origins of the idea that symbols are in the head are to be found in the Cartesian idea of what it is to be a thinking thing (Fetzer 1996). Dreyfus's summary, reported in Hutchins (1996, p. 357), provides as succinct an account of the origins as one might wish for:

> GOFAI is based on the Cartesian idea that all understanding consists in forming and using appropriate symbolic representations. For Descartes, these representations were complex descriptions built up of primitive ideas or elements. Kant added the important idea that all concepts are rules for relating such elements, and Frege showed that the rules could be formalized so that they could be manipulated without intuition or interpretation.

Thus entities thought to be inside the head are modeled on a class of entities external to it, symbolic representations. Symbolic logic in turn was essential to early computation, and it was an easy step to conceive of computers as somehow representing a person's reasoning processes, and to see a person as resembling a machine in some sense. If the brain could be considered as a digital machine, then such a machine — the Turing universal machine — could compute any function provided it was explicitly specified, and it could be programmed to do so.

What has happened according to Hutchins' (1996) alternative history of cognitive science is that the actual person, such as Turing, the inventor of the universal machine, dropped out of the picture. After all, it was originally he who physically manipulated those symbols in interacting with the world whose cognitive properties while doing the manipulation were not the same as the properties of the symbol systems which were manipulated (see also Clark 1997, Chapter 3). The manipulation of the system by the cognitive properties of the human produces a computation but it does not follow that this computation is happening in the person's head.

Hutchins' navigation example makes this point well. This system is in large part characterized by the execution of many formal operations, but not all the formal operations used to process the computational properties of this system are in the heads of the quartermasters. Quite a few are in the material environment which is shared and produced by them in their mutual interactions. The fact that this could be bleached out of the history of cognitive science is due to the central assumption that the formal manipulation of abstract symbols is both a necessary and sufficient condition for modeling human cognition. What has really been modeled here according to Hutchins is the operation of a sociocultural system from which the human actor has been deleted since what does the work in this model is the interaction of abstract symbol systems and not any one person. The computer then was hardly made in the image of a human being; it was made in the image of the formal manipulations of abstract symbols. It is in this way that symbols have found their way into human heads, by stipulating an isomorphism between symbol structures and neuronal activity patterns where the latter are said to bear 'a one-to-one relation to the symbol structure of Category 4 in the corresponding program [Category 4 = data stored in the computer]' (Vera and Simon 1993c, pp. 120–121).

As a result, there is a vast gulf between internal cognition and the outside world of experience of people perceiving and moving about in their changing environments, as is recognized and misidentified as a left-over problem for the physical symbol system hypothesis. It was thus entirely logical that the testing of cognition happened in the laboratory, and that the tasks to which subjects were subjected were at the unusual end of the spectrum of human cognitive activity, such as puzzle solving or playing chess. These are cultural activities which do not belong to the humdrum cognitive activities of everyday life, and studying these yielded results which are not typical or representative of ordinary human cognition.

Given this analysis, it can be concluded that the empirical success the physical symbol system hypothesis has apparently enjoyed is rather a case of confirmation of an abstract model of symbol manipulation embedded in larger cultural production than confirmation of the symbol processing nature of a single human being.

What emerges from the preceding is an indication of how cognition is spread or distributed in the world. In one sense we have always known this. Organizations, including schools, are examples of such successful distribution of human cognitive activity since no one individual can carry out all the cognitive functions conducted under the roof of one school or factory. Many computational tasks are too complex for one individual mind (Clark 1997) and, as a consequence, the mind/brain off-loads some important problem-solving tasks onto external structures. Such external structures in turn exhibit their own dynamics which react back on and shape individual cognition in reciprocal interaction. An example is living in a democracy with its established rules and freedoms as opposed to living in an authoritarian society in which freedom of speech and freedom of association may be curtailed. These structural, legal, and social conditions shape how the citizens behave and exercise whatever rights they possess, including the freedom to learn. Since humans have limited cognitive capacities, the contracting out of cognitive tasks, or the cognitive division of labor, has been essential in order for people to become the kinds of advanced knowers that they are by some kind of cognitive bootstrapping. It is in this sense that external structures are extensions of the mind. This is rarely considered explicitly, and yet it is deeply embedded in everyday life. Patients who suffer from advanced Alzheimer's disease (Clark's example) depend on external structures to find their way around in their world. Without markers on doors, shopping-lists stuck to the refrigerator, a favorite chair in the right place, or the milk bottle by the back-door, these people lose their place in the world. Shifting objects, or altering anything in the environment serves to destabilize them and quite literally to rob them of their minds. Mind and world are indeed indivisible, as Lave had argued earlier.

Translated into larger social contexts, it can be seen that organizations serve a similar purpose as having a favorite chair in its normal position. By means of their structures, practices and norms they provide external resources which define the contexts for people being in them. This applies also to linguistic structures such as policies and regulations which help reshape some cognitive tasks into formats which may be better suited to human computational capacities. The role and function of language goes far beyond that of communication. It is reasonable to see it as presenting a kind of trade-off between 'culturally achieved representations and what would otherwise be ... time-intensive and labor-intensive internal computation.' (Clark 1997, p. 200). This includes the off loading of memory by way of written texts; the devising of labels which serve as a kind of cognitive shorthand in navigating the environment successfully; language which allows people to coordinate their actions with those of others; and it also allows people to manipulate their own attention by inner rehearsal of speech.

Given these various functions, it may be concluded with Churchland (1995, p. 270) that 'language makes it possible, at any time, for human cognition to be *collective.*' Language as the cultural externalization of individual thought transcends the cognitive system of the individual, and any single individual's life span, and conjoins it with that of others. The result is a far more powerful system with vastly increased cognitive resources to solve shared problems. The full import of this can only be appreciated on the background of human learning capacity, which seems to be characterized by extreme path dependence (Elman 1994). This means that neural net learning, given the specific learning algorithm, the back propagation of error, is dependent on the sequence in which training cases are presented. Extrapolating this to human learning, some ideas have to be in place before others can be learnt, and given that a net's knowledge is located in its connection weights, any new learning is similarly constrained by those weights in so far as they serve as the original parameters for new learning. It is thus easy to see that neural net learning cannot jump around (Clark 1997) to any place it pleases. If this is so, the role of public language, in allowing the packaging and migration of ideas between individuals (Clark 1997, p. 205), truly collectivizes human cognition and in so doing makes available immensely powerful cognitive resources way beyond individual capacity.

## Socially Distributed Cognition in Practice: Some Implications

In light of this account of the history of cognitive science with its emphasis on studying cognition in the confines of the laboratory, it is clearly important to support *situated action* in its conduct of ethnographic field studies as a first descriptive step to understand ordinary cognitive activity in its specificity. The contribution *situated action* makes is in terms of displaying the minutiae of everyday life, fine-grained descriptions whose frameworks are informed by such ethnographic concepts as 'normal routines of collaborative work, distributions of accountability and authority, mutual availability of and attention to information sources, mutual understanding in conversation, and other characteristics of interaction that are relevant to the functional success of the participants' activities' (Greeno 1997, p. 7). This approach enables researchers to focus on how human actors actually display their cognition in interaction with others and with material-structural systems which in turn shape and change their cognitive behavior. In doing so, this direction is not only compatible with evolution and biology in so far as it allows for human perception and locomotion, it does so within the context of socially and culturally created systems. It thus coheres much better than the physical symbol system hypothesis with what is known about human origins, biological, physiological, social and historical.

What it seems to lack, on the other hand, is a causal account of the actual mechanisms by means of which actors accomplish what they do. While cognition is social in the sense of being accomplished within broader cultural frameworks, it is nevertheless also in some sense the accomplishment of an individual mind/brain in interaction with other minds/brains. While the symbol processing view located all cognition inside the head and treated the environment as a mere extension of memory, *situated action* appears to locate it all outside in the interactions between people and social structures. Yet short of attributing causal powers to structures which would leave open the question as to how such powers came about, it is humans who interpret and make sense of the structures and their interactions with them and with one another. However, *situated action* remains agnostic about the fine-grained physiological detail of how humans can do what they do so well, including the creation of human history and culture. In addition, it clearly is the case that humans do manipulate symbols successfully, and that linguistic structures have a role to play in the human cognitive economy, a facet not explained by *situated action* although acknowledged.

It is against the background of the preceding discussion that our views of socially distributed cognition and practice in education and educational administration are developed in the next chapters. We follow established practice by pointing out both what we have learnt from, and where we agree with, other perspectives. Needless to say, we also state as concisely as possible where our own approach differs and where it develops new directions for theorizing and practice. This will become especially clear in our last chapter where we spell out some implications of socially distributed cognition for the conduct of research in education and educational administration.

## References

Anderson J.R. (1983). *The Architecture of Complexity*. (Cambridge, MA: Harvard University Press).

Anderson J.R., Reder L.M. and Simon H.A. (1996). Situated learning and education, *Educational Researcher*, 25(4), pp. 5–11.

Anderson J.R., Reder L.M. and Simon H.A. (1997). Situative versus cognitive perspectives: Form versus substance, *Educational Researcher*, 26(1), pp. 18–21.

Bechtel W. (1990). Connectionism and the philosophy of mind: An overview, in W. G. Lycan (ed.) *Mind and Cognition — A Reader*. (Oxford: Basil Blackwell).

Bereiter C. (1991). Implications of connectionism for thinking about rules, *Educational Researcher*, 20(3), pp. 10–16.

Chaiklin S. and Lave J. (eds.) (1993). *Understanding Practice — Perspectives on Activity and Context*. (Cambridge: Cambridge University Press).

Churchland P.M. (1995). *The Engine of Reason, The Seat of the Soul*. (Cambridge, MA: MIT Press).

Churchland P.S. and Sejnowski T.J. (1990). Neural representation and neural computation, in W.G. Lycan (ed.) *Mind and Cognition — A Reader*. (Oxford: Basil Blackwell).

Churchland P.S. and Sejnowski T.J. (1994). *The Computational Brain*. (Cambridge, MA: MIT Press).

Clancey W.J. (1993). Situated action: A neuropsychological interpretation, response to Vera and Simon, *Cognitive Science*, **17**, pp. 87–116.

Clark A. (1997). *Being There: Putting Brain, Body, and World Together Again*. (Cambridge, MA: MIT Press).

Elman J. (1994). Learning and development in neural networks: The importance of starting small, *Cognition*, **48**, pp. 71–99.

Evers C.W. (1991). Towards a coherentist theory of validity, in G. Lakomski (ed.) *Beyond Paradigms: Coherentism and Holism in Educational Research*, International Journal of Educational Research, Special Issue, **15**(6), pp. 521–535.

Evers C.W. and Lakomski G. (1991). *Knowing Educational Administration*. (Oxford: Elsevier).

Evers C.W. and Lakomski G. (1996). *Exploring Educational Administration*. (Oxford: Elsevier).

Fetzer J.H. (1996). *Philosophy and Cognitive Science*. (New York: Paragon House).

Greeno J.G. (1997). On claims that answer the wrong questions, *Educational Researcher*, **26**(1), pp. 5–17.

Greeno J.G. and Moore J.L. (1993). Situativity and symbols: Response to Vera and Simon, *Cognitive Science*, **17**, pp. 49–59.

Haugeland J. (1985). *Artificial Intelligence: The Very Idea*. (Cambridge, MA: MIT Press).

Hastie R. and Pennington N. (1991). Cognitive and social processes in decision making, in L.B. Resnick, J.L. Levine and S.D. Teasley (eds.) *Perspectives on Socially Shared Cognition*. (Washington, DC: American Psychological Association).

Hutchins E. (1991). The social organization of distributed cognition, in L.B. Resnick, J.L. Levine and S.D. Teasley (eds.) *Perspectives on Socially Shared Cognition*. (Washington, DC: American Psychological Association).

Hutchins E. (1996). *Cognition in the Wild*. (Cambridge, MA: MIT Press).

Johnson-Laird P.N. (1983). *Mental Models*. (Cambridge, MA: Harvard University Press).

Lakomski G. (ed.) (1991). *Beyond Paradigms: Coherentism and Holism in Educational Research*, International Journal of Educational Research, Special Issue, **15**(6), pp. 499–597.

Lave J. (1988). *Cognition in Practice*. (Cambridge: Cambridge University Press).

Lave J. (1991). Situating learning in communities of practice, in L.B. Resnick, J.L. Levine and S.D. Teasley (eds.) *Perspectives on Socially Shared Cognition*. (Washington, DC: American Psychological Association).

Lave J. and Wenger E. (1991). *Situated Learning*. (Cambridge: Cambridge University Press).

Levine J.M. and Moreland R.L. (1991). Culture and socialization in work groups, in L.B. Resnick, J.L. Levine and S.D. Teasley (eds.) *Perspectives on Socially Shared Cognition*. (Washington, DC: American Psychological Association).

Lycan W.G. (ed.) (1990). *Mind and Cognition, A Reader*. (Oxford: Blackwell).

Newell A. and Simon H.A. (1972). *Human Problem Solving*. (Prentice-Hall).

Newell A. and Simon H.A. (1976). Computer science as empirical enquiry: symbols and search, in M.A. Boden (ed.) (1990). *The Philosophy of Artificial Intelligence*. (Oxford: Oxford University Press).

Newell A., Rosenbloom P.S. and Laird J.E. (1996). Symbolic architectures for cognition, in M.I. Posner (ed.) *Foundations of Cognitive Science*. (Cambridge, MA: MIT Press).

Norman D.A. (1993). Cognition in the head and in the world: An introduction to the special issue on situated action, *Cognitive Science*, **17**, pp. 1–6.

Packer M.J. and Winne P.H. (1994). The place of cognition in explanations of teaching: A dialog of interpretive and cognitive approaches, *Teaching and Teacher Education*, **10**(1), pp. 1–21.

Posner M.I. (ed.) (1996). *Foundations of Cognitive Science*. (Cambridge, MA: MIT Press).

Resnick L.B., Levine J.M., and Teasley S.D. (eds.) (1991). *Perspectives on Socially Shared Cognition*. (Washington, DC: American Psychological Association).

Rogoff B. and Lave J. (eds.) (1984). *Everyday Cognition: Its Development in Social Context*. (Cambridge, MA: Harvard University Press).

Rumelhart D.E. and McClelland J.L. (eds.) (1986). *Parallel Distributed Processing*. Volumes I & II. (Cambridge, MA: MIT Press).

Scribner S. (1984). Studying working intelligence, in B. Rogoff and J. Lave (eds.) *Everyday Cognition: Its Development in Social Context*. (Cambridge, MA: Harvard University Press).

Shulman L.S. (1987). Knowledge and teaching: Foundations of the new reform, *Harvard Educational Review*, **57**(1), pp. 1–22.

Simon H.A. (1979). *The Sciences of the Artificial*. (Cambridge, MA: MIT Press, 2nd Edition).

Simon H.A. (1990). Invariants of human behavior, *Annual Review of Psychology*, **41**, pp. 1–19.

Smolensky P. (1988). On the proper treatment of connectionism, in D.J. Cole *et al.* (eds.) *Philosophy, Mind, and Cognitive Inquiry*. (Netherlands: Kluwer).

Tienson J.L. (1990). An introduction to connectionism, in J.L. Garfield (ed.) *Foundations of Cognitive Science: The Essential Readings*. (New York: Paragon House).

Vera A.H. and Simon H.A. (1993a). Situated action: A symbolic interpretation, *Cognitive Science*, **17**, pp. 7–48.
Vera A.H. and Simon H.A. (1993b). Situated action: Reply to William Clancey, *Cognitive Science*, **17**, pp. 117–133.
Vera A.H. and Simon H.A. (1993c). Situated action: Reply to Reviewers, *Cognitive Science*, **17**, pp. 77–86.

# PART II

# Administrative Practice

# 4

# Leadership in Context

In Part I of this book, we spent quite a bit of time explaining our general framework for understanding the representation of practical knowledge both for individuals as well as for human collectivities who interact with, and are in turn shaped by, their social-material environments. In the second, much larger part, we offer a range of applications designed to show how our theory of practice works in respect of select administrative and educational practices. Specifically, the aim is to demonstrate how our coherentist theory of practice constrains options for understanding such central concerns as organizational design, decision-making, ethical conduct, administrator training, and leadership. It is the latter issue which is the topic of this chapter.

It is hardly controversial to observe that Leadership-as-a-good-thing is deeply entrenched in our common culture. Much is expected of leaders and leadership when economic, managerial, or other crises have to be met. The solutions to restructuring for purposes of greater efficiency and effectiveness, whether in private or public sector organizations, are widely sought in better leadership or 'strong leaders' who are believed capable of steering the organization in desired directions.

While there is no doubt that there have been, and are, strong individuals who by virtue of their abilities and personalities were and are able to have a positive impact on organizations, the concept of leadership has acquired a privileged status which seems to have removed it from critical purview. By this we mean that leadership is commonly, and apparently universally, accepted as an essential human quality. Such pervasive acceptance is not at all threatened by the critical literature of leadership, which is voluminous, but the debates in the field mainly proceed on the assumption of the unquestioned essence of leadership as a real empirical phenomenon, which somehow has to be captured. Once this has been achieved, it is presumed that we then might know how to *create* good leaders.

In this chapter we raise some doubts about the purported essence of leadership, the claims advanced in terms of its efficacy and scope, and the methodology used by researchers of leadership. The line of argument to be developed is that leadership — whether in its traditional or *New Leadership* conceptions — makes assumptions about human cognition, which cannot in principle explain either individual or organizational practice. Its empiricist theoretical framework is too narrow to capture what is believed to be the essence of good (or transformational) leadership: the ability to motivate others, create a vision for the future, make appropriate judgments in the face of uncertainty, and enthuse followers to work 'beyond expectation'.

We have organized our comments around a number of key theses, which will indicate the flavor of the argument. It is fair to indicate that these issues provide the mere outline of a longer term research agenda, which is still evolving. The present chapter is an initial exploration of some of the more telling difficulties, which beset this so popular field of study and practice.

Central Theses:

1. The concept of Leadership is without a referent. There is no natural object or kind in nature to which Leadership refers. Much current usage is *essentialist*.
2. As a folk psychological and functionalist concept, Leadership is massively disconnected from causation.
3. The various findings of descriptive-quantitative (and qualitative) Leadership studies, employing instruments such as the LBDQ, are artifacts of methodology rather than scientific accounts of empirical phenomena.
4. The functionalist framework of most Leadership studies, methodologically supported by hypothetico-deductivism, inappropriately sums specific, context-dependent results across all organizations regardless of difference. It thus fails to account for specificity of context and practice.
5. Organizational practice is always interpreted practice. Interpretations of Leadership are context-dependent, specific, and thus not universalizable. The concept of Leadership *fragments* at the local level.
6. As linguistic abstractions (sentential representations) from specific action contexts, Leadership theories systematically fail to account for organizational practice — the *how* of Leadership whose representational structure is a matter of neuronal not sentential organization.
7. From a *naturalistic-coherentist* perspective, there may be as many different accounts of leadership as there are organizational contexts. Law-like statements about leadership, as postulated by empiricist theories of leadership, are not to be had. Generality, insofar as it can be obtained, would be a matter of the coherence of accounts in a specified context. We may develop modular rather than system-like accounts.
8. Effective practice causally depends on the activation of appropriate neuronal patterns of Leader and followers. Since these do not follow hierarchical structures, the potential for effective practice resides throughout the organization.
9. Organizational learning is thus the key to effective administrative practice with the consequence of creating appropriate web-like organizational structures which maximize the local production of knowledge and facilitate the correction of error through feedback mechanisms.

(This point is developed more fully in Chapter 5, this volume).

## Leadership and Effectiveness: The Current State of Play in Educational Administration Research

Hallinger and Heck (1996, p. 5) begin their recent review of empirical research about the principal's role in school effectiveness with the following words: 'The belief that principals have an impact on schools is long-standing in the folk wisdom of American educational history. Studies conducted in recent decades lend empirical support to lay wisdom'. We might add that such folk wisdom is not the prerogative of American studies alone since the role of the principal in terms of effecting change is equally strongly held in the UK and in Australian research. The authors add in their opening paragraphs that 'this relationship is complex and not easily subject to empirical verification' (Hallinger and Heck 1996, p. 6), an assessment shared by, amongst many others, the reviews conducted by Bridges (1982), Bossert *et al.* (1982), and Howe (1994).

Hallinger and Heck's reassessment is important both in terms of their conclusions and in terms of presenting the kinds of conceptual approaches, which have characterized research in the field of educational leadership and effectiveness since 1980. The reason why conclusions and presentation of research models are important is that they are indicative of a stagnant research program, which is markedly under-theorized, as will be shown in the following. This strong conclusion, of course, is ours and not the authors'. Their own criticisms are softened by a belief that some progress can be noted insofar as greater emphasis has been placed, conceptually and empirically, on the complex interplay of (school) internal, environmental and personal characteristics of the principal. In their view, '... the principal's role is best conceived as part of a web of environmental, personal and in-school relationships that combine to influence organizational outcomes' (Hallinger and Heck 1996, p. 6). However, they also concede that at present '... the specific nature of these complex interactions across sets of variables within a model of principal effectiveness remains unclear' (Hallinger and Heck 1996, p. 38). Practically all the studies they analyzed used cross-sectional and correlational designs with surveys and interviews as the preferred data collection techniques. In the authors' view, such nonexperimental studies are less well equipped to draw causal inferences than are experimental studies since independent variables are not manipulated. Determination of causation is thus much more difficult because all relevant independent variables must be controlled which, in a theoretically weak model, may not be specifiable or specified, and thus elude control and create a major threat to validity.

A related and most important concern is that any interpretation of the complexity of the relationships depends on the sophistication of the theoretical model itself. If the model is overly simplistic, then analyses may be simplistic. In this case, results are ambiguous or lack validity leading the authors to conclude that '[i]n the absence of an explicated theoretical model, the researcher often cannot be sure what has been found' (Hallinger and Heck 1996, p. 17). A further issue is the analytical techniques used themselves. More rigorous techniques would lead to stronger conclusions than the ones reached in the studies examined.

As for models or conceptualizations of the principal's role in school effectiveness, Hallinger and Heck adapted a classification scheme originally developed by Pitner (1988) which needs to be outlined just briefly. These models are defined as (a) *direct-effects* (Model A), (b) *mediated-effects* (Model B), (c) *antecedent-effects* (Model A1, B1), and (d) *reciprocal-effects* (Model C).

The *direct-effects* model tests the principal's effects on school outcomes directly without intervening variables, is very common in the literature but is also no longer considered useful because of its 'black box' approach: an empirical relation is tested without having any knowledge about the process of achieving an impact. Leadership itself remains unexplained, and the purported impact remains a mystery.

The *mediated-effects* model assumes that whatever impact the principal has is achieved by interaction with, or manipulation of, organizational features of the school. Leadership thus works through others.

The *antecedent-effects* model postulates that the principal's role may be both a dependent or independent variable. This means simply that principals influence, and are themselves influenced by, other variables in the school environment. As a result the principal's actions may be seen as both the outcome of direct or mediated effects. In the authors' view, this model presents an advance because it offers a more comprehensive picture of the principal's role in school effectiveness.

The *reciprocal-effects* model emphasizes the interactive relationship between the principal and organizational features of the school and its environment. Principals learn to adapt to the organization, in which they work and subsequently change their behavior, and presumably also the impact they have on the school's effectiveness. Leadership is considered as an adaptive process rather than a unitary and fixed feature.

The conclusions Hallinger and Heck draw are illuminating. They believe that unlike Bridges' scathing comments 'that studies of school administrators are intellectually random events' (Bridges 1982, p. 22) there has been a *conceptual* advance in that 'virtually all of the studies could be classified as *theoretically* informed' (Hallinger and Heck 1996, p. 33). They mean by this that researchers defined their constructs and gave reasons for their choice of variables. However, there is also an advance in terms of explicitly linking studies to theoretical positions both in terms of the relationship of variables to those positions and in terms of the relationship of the leadership construct to a broader theoretical framework.

On the methodological side there is also progress to report, especially in terms of applying more sophisticated analytical tools, such as more powerful versions of structural modeling, which were more appropriate to the theoretical orientations proposed. However, the interpretation of data generated by correlational studies is still limited because of an absence of longitudinal research of both quantitative and qualitative kinds. Hallinger and Heck also note the emergence, since the 1980s, of new leadership constructs potentially useful in explicating leadership effects such as instructional and transformational leadership, as well as models of leadership inspired by the work of Bolman and Deal and Sergiovanni.

As for that most important question: Do principals make a difference, the authors advise that 'considerable caution' needs to be applied to the results of the studies they examined. Their qualifications are important to note:

1.  Theoretical model type made a difference in what was found; the more sophisticated models (Model B, B1) showed more positive albeit still weak relation between leadership and effectiveness;
2.  While leadership can make a difference, attention must be paid to context, to the conditions under which the effect is achieved, i.e. socioeconomic environment; but studies are too disparate in their ideas of leadership and context variables to be able to specify the relevant contingencies;
3.  Where there was a positive difference found it related more to internal school processes, which were in turn linked to student learning;
4.  Studies with positive findings consistently show up goal orientation as a significant factor which, however, is also influenced by environmental factors. Hence this finding needs to be qualified as well.

Hallinger and Heck conclude that one need not be unduly pessimistic about this motley array of qualified research results since much has been gained by acknowledging positive indirect leadership effects, mediated through in-school variables. This, they claim, in no way diminishes the principal's importance since 'achieving results through others is the essence of leadership' (Hallinger and Heck 1996, p. 39).

However, and although they couch their conclusions in more placatory language, they do agree with Bridges' (1982, p. 25) (as well as Bossert's) conclusions: ' ... there is no compelling evidence to suggest that a major theoretical issue or practical problem relating to school administrators has been resolved by those toiling in the intellectual vineyards since 1967'. Bridges drew a further conclusion from his investigation, which is worth repeating:

> If the intellectual sterility and marginal utility of this work is characteristic of research in educational administration, the profession is in difficulty. Studies of 'no significance' are patently more troubling than studies with 'no significant difference' (Bridges 1982, p. 17).

## Leadership and Effectiveness: Early Conceptual and Methodological Criticisms

The old adage that those who don't know their history are condemned to repeat it, is only too true of the history of research on leadership and effectiveness. In particular, the absence of any knowledge of the parent discipline of organizational theory, and especially the debates surrounding leadership and effectiveness which arose in the wake of the Human Relations movement which made leadership the most prominent organizational theory concept, undermines the intellectual base of the education debate in the field. This is not the place to rehearse the history of the Human Relations movement, nor to examine the enormous bulk of empirical studies conducted, but to draw attention to perennial problems as they were already visible and acknowledged at the origins of researchers' concern with the phenomenon of leadership. These worries continue, and are clearly evident in Hallinger and Heck's reassessment of leadership and effectiveness in educational administration.

The importance of leadership, as evidenced in a truly voluminous literature, has apparently never been in doubt in terms of being able to shape and give direction to social organization. Hemphill (1949, p. 3), an early advocate, made the case in favor of leadership quite clear when he noted that 'Both laymen and scientists agree that if we can understand the selection and training of leaders we can begin to take adaptive steps toward controlling our own social fate'.

Heavily influenced by social psychology, the study of leadership at the beginning of this century first concentrated on identifying a unitary trait or personal characteristic that would clearly mark a person as a leader. No such trait has been found, and the *unitary trait* theory was replaced by a *constellation-of-traits theory* (Gibb 1959, p. 914). Although this theory permitted a pattern of traits which could differ between leaders and situations, the 'why' of leadership was still conceived of as a function of personality (e.g. Stogdill 1948) and is thus a mere variant of the former unitary trait theory. Because of the difficulties of identifying leaders — people formally or informally designated as such are not necessarily 'leaders' — it has been suggested that that person is the leader who exerts 'influence' over group members where both influence and its direction are agreed upon by the group members/ followers. However, a more objective approach still was deemed to be attention to leader behavior occurring in a group. 'Leadership acts may then be defined as the investigator wishes, and leaders are to be identified by the relative frequency with which they engage in such acts' (Gibb 1959, p. 916). Thus, while personality characteristics continued to be important, the emphasis had shifted to the impact of leadership on groups' performance or satisfaction.

Underlying the focus on leadership since the human relations school and the Hawthorne studies is what Bowers and Seashore (1973, p. 445) describe as a 'commonly accepted theorem':

> Leadership in a work situation has been judged to be important because of its connection, to some extent assumed and to some extent demonstrated, to organizational effectiveness. Effectiveness, moreover, although it has been operationalized in a variety of ways, has often been assumed to be a unitary characteristic. These assumptions define a commonly accepted theorem that leadership ... is always salutary in its effect and that it always enhances effectiveness.

Indeed, following Perrow's (1986, p. 85) classification, the human relations school has two branches. The first is concerned with morale and productivity and is by far the most empirically researched, while the second, still closely related branch, is more interested in the structuring of groups. A basic premise, especially of the first, is as follows:

'Good leadership' is generally described as democratic rather than authoritarian, employee-centered rather than production-centered, concerned with human relations rather than with bureaucratic rules, and so on. It is hypothesized that good leadership will lead to high morale, and high morale will lead to increased effort, resulting in higher production.

While these assumptions seem so eminently commonsensical and true, Perrow sums up his investigation by stating that forty years of consolidated research only managed to find that human behavior is far too complex to allow any simple kinds of conclusions to be drawn which characterized the hopes of the Human Relations theoreticians. Even more importantly, Perrow claims that all empirical studies conducted on the relationship between attitude and performance simply set out to prove that 'happy employees are productive employees' and that the findings were not robust enough to support the assumptions. In fact, the presumed direction of causality could be reversed, as was argued by Lawler and Porter (1967) in their important analysis of thirty empirical studies. In other words: high productivity creates high satisfaction.

Although Lawler and Porter agree that there is a low but consistent relationship between these two variables, they note that it is not all clear *why* the relationship exists. This raises the problem of whether job satisfaction is indeed important, and if so, whether organizations should take steps to maximize it. In their view, it seems organizationally more prudent to reward high performance by satisfying employees' higher needs, that is, provide them with more autonomy and avenues for self-actualization, which, in turn, has positive outcomes in terms of lower absenteeism and turnover which are positively related to productivity. Of importance is thus the relationship between the two variables, and not just 'satisfaction' as a single feature, as considered in the human relations mode. Furthermore, however, as Perrow (1986, p. 87) notes, some jobs leave no room for high performance, and productivity depends more on technological changes or economies of scale than human effort. (According to Lawler and Porter (1967), there were few studies reported in the literature regarding the relationship between satisfaction and performance post–1955).

The study of the effects of leadership behavior/style on group performance began in earnest in about 1945 and is commonly identified with the Ohio State Leadership Studies. They left a far-reaching legacy which still characterizes much contemporary educational leadership research: the development of the Leader Behavior Description Questionnaire (LBDQ) (see Hemphill and Coons 1973, pp. 6–38), the Leadership Opinion Questionnaire (LOQ), a leader self-assessment instrument, and the two-factor theory of leadership described by the concepts of (1) 'consideration' ($C$), and (2) 'initiating structure' ($S$) (Halpin and Winer 1973, and see also the studies reported in Stogdill and Coons 1957). Developed in interdisciplinary discussions with psychologists, economists and sociologists, the LBDQ was designed to 'be adaptable to studies in widely different frames of reference. This would make it possible to include such an instrument in each individual research design, thereby contributing to an integration of research findings that would not be possible otherwise (Hemphill and Coons 1973, p. 7). Arrived at through various factor analyses, the two remaining factors 'consideration' and 'initiating structure', are similarly entrenched in organizational-administrative folklore. $C$ describes leader behavior which is warm, shows mutual trust, respect and friendship while $S$ refers to leader behavior which 'organizes and defines relationships or roles, and establishes well-defined patterns of organization, channels of communication, and ways of getting jobs done' (Bowers and Seashore 1973, p. 442). Unlike the earlier Human Relations view, according to the Ohio Studies, these two factors were not seen as opposites — leaders who are either good on human relations or managed to get jobs done — but both appeared to be equally important. Despite their immense popularity in industrial psychology, management and organization theory, where they have become articles of faith, 'consideration' and 'initiating structure' are variables whose predictive powers regarding organizational or group effectiveness are simply not proven (Korman 1966, p. 360).

There is no clear support for the background assumption that leaders high on $C$, for example, 'cause' better performance in subordinates; rather, Korman notes, causation might well be the other way round: leaders may show more consideration toward already well-performing subordinates, or they may show more $C$ to those who support them. Alternately, supervisors might be higher on $S$ in case of low performing groups. All these possibilities need to be tested experimentally, but have not been. Furthermore, supervisor ratings might be affected by variation in organizational size and climate. Importantly, however, Korman (1966, p. 355) argues that we need to be able to provide 'a systematic conceptualization of situational variance as it might relate to leadership behavior ...' rather than merely acknowledge its importance, as had been the case in the studies he analyzed.

Amongst the various attempts to broaden this two-factor theory of leadership, Fiedler's 'contingency' theory (Fiedler 1967) shall be mentioned briefly because it introduces considerably more complexity by adding 'group climate' as a central feature believed to affect leadership effectiveness. Fiedler's (1973, p. 468) model postulates that 'the performance of interacting groups is contingent upon the interaction of leadership style and situational favorableness.' He thus reinforces the notion that group effectiveness is a feature of leader attributes as well as situational factors. He claims that task-oriented leaders do well in very favorable as well as very unfavorable situations while people-oriented leaders do better in situations of intermediate favorableness. The predictor measure used is his least-preferred co-worker score (LPC). Suffice it to say that while Fiedler has gathered a lot of supporting data, there are more complexities involved in terms of defining the nonleadership variables, that is, 'situational favorableness' changes, and so do interpersonal perceptions of 'followers' and leaders. Yet the important insight derived from this melding of leader characteristics and situational factors is, as Hemphill (1949, p. 225) put it, that ' ... there are no absolute leaders, since successful leadership must always take into account the specific requirements imposed by the nature of the group which is to be led, requirements as diverse in nature and degree as are the organizations in which persons band together'.

The sheer complexity and contingency of these factors led Perrow (1986, p. 92) to state that

> If leadership techniques must change with every change in group personnel, task, timing, experience, and so on, then either leaders or jobs must constantly change, and this will make predictions difficult. At the extremes, we can be fairly confident in identifying good or bad leaders; but for most situations we will probably have little to say. We may learn a great deal about interpersonal relations but not much about organizations.

## Leadership and Effectiveness: A Regressive Research Program

So what have we learnt from looking back at the origins of leadership studies? The kinds of problems and results Hallinger and Heck report in their latest assessment of leadership and effectiveness in the context of school leaders have typically been reported in the parent discipline — a good twenty years earlier, if not more. But there seems to be no recognition of that history, and the current state of conceptualizing leadership as evidenced in the empirical work reported, is as limited as that reported and variously critiqued in the field of organizational studies. The question which has fascinated and motivated researchers from the beginning, whether leaders make a difference or not, has to date not been answered satisfactorily. The best we can say — on the basis of the empirical studies carried out — is that we think so, somehow!

The major point to make here is that our preoccupation with leadership and its impact while understandable in terms of its commonsense and possibly political, appeal — is a preoccupation which we should not pursue in this form because it is *epistemically unproductive.*

There are specifically two points, which emerge from the accumulated history of leadership research which, for present purposes, are important. The first is the inconclusiveness of empirical research results, and the second, the oft-repeated observation that specific requirements and the diverse natures and degrees of organizations studied makes a difference in determining leadership effects; the importance of situational 'contingent' factors, as they arise in the empirical studies, seem to point to the fact that there is no *essence* to leadership, that leadership means different things to different people in different contexts. In other words, the 'why' of leadership remains a mystery, and this is not surprising given the empiricist hypothetico-deductive framework within which the bulk of leadership studies were conducted.

Consider the structure of a typical study, suggested by Bowers and Seashore (1973, p. 447) as more appropriate than the earlier behavioral ones. What the authors suggest has become standard fare also in studies of leadership in education. It contains the following features:

> (a) Measures reflecting a theoretically meaningful conceptual structure of leadership, (b) an integrated set of systematically derived criteria; and (c) a

treatment of these data, which takes account of the multiplicity of relationships and investigates the adequacy of leadership characteristics in predicting effectiveness variables (Bowers and Seashore 1973, p. 447).

'A theoretically meaningful conceptual structure of leadership' denotes a *leadership construct* made up of whatever variables the researcher hypothesizes as important. For present purposes, the structure and associated difficulties of such a hypothetico-deductive model can be shown as follows.

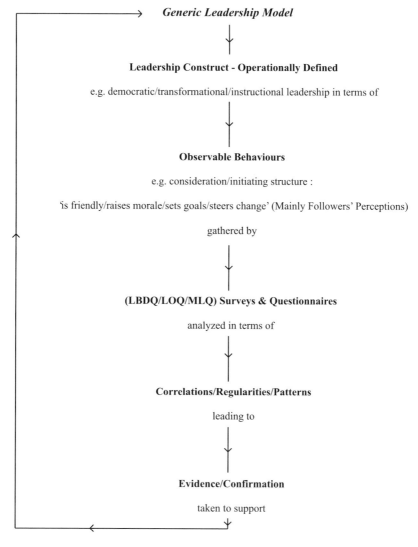

**Figure 4.1** Deductive Leadership Model

As a central feature of *logical empiricism*, this hypothetico-deductive account of scientific theory and practice, no longer accepted as valid in philosophy of science and epistemology, postulates that empirical evidence consists of singular observations, i.e. observation reports of behaviors which are hypothesized to be representative of whatever construct is being tested. These individual claims (observation reports) are at the bottom of the hierarchy, they form the *foundations* for claims to leadership at the top of the hierarchy. These claims are believed to be empirically testable, and the process of deduction and testing makes up the so-called hypothetico-deductive account (see Evers 1995, p. 2; and Evers and Lakomski 1991, 1996 for more detailed examination). In addition, an important feature of this empiricist (positivist) account of scientific theory is the notion of *operational definition*. Since empiricist theory prescribes that its foundation is based on observation reports, every claim made — whether about 'observable' or 'theoretical' entities — has to be amenable to empirical definition (e.g. Fiedler's account). Now this appears to be straightforward with regard to observable physical objects, but seems problematic in relation to non-observable, theoretical entities, like 'leadership', for example, or any type of value such as 'good citizenship', 'equality' or 'justice'. The way out for logical empiricists was to operationalize them, that is, to develop some measurement procedure which would 'capture' the elusive entity in the absence of the possibility of direct observation. Operational definitions played a large part in the traditional scientific account of leadership, as seen in the discussion above, as well as in educational administration (see the Theory Movement), and is still found in the work of Hoy and Miskel (1996).

The history of leadership studies, at least in the early empiricist tradition, seems to be driven by the ongoing effort to find more appropriate factors/variables which can be taken as representative of its true nature: from single trait to multiple traits, to increasingly complex postulated interrelationships between such things as organizational and group structure and environmental factors, commonly subsumed under the umbrella term of the 'situational factor'. Operationalizing these variables by means of designated observable administrative behaviors reported in surveys or questionnaires such as the LBDQ, and gathering and analyzing these via increasingly complex statistical-quantitative methodology was believed to provide appropriate empirical support which would then lead to proper generalizations about leadership across organizational contexts. If that could be achieved, leaders could be trained and organizations be made more efficient.

It seems to be assumed, without argument, that whatever leadership constructs are composed, that they do refer to a phenomenon in nature, and that by re-conceptualizing the constructs/theories, we get closer to its essence. On the face of it, why should we not presume that there is such as thing as leadership since leaders are ubiquitous amongst baboons, birds and bees, just as they are amongst other card-carrying members of the animal kingdom: humans. So where is the problem? Just because we have a vocabulary, or conceptions, that is, linguistic representations of leadership, it does not follow that there really is such a thing in nature to which they refer. Following the argument put by Churchland (1993, p. 284), it is more than likely that all of our commonsense frameworks may be 'unconnected to the world by way of reference of its singular terms and the extension of its general terms.'

This point is easily appreciated when we consider that it is possible to employ a false scientific theory which guides our observations, such as 'caloric fluid' or 'phlogiston', or 'ether'. We can express our observations in the theory's terminology although we know it to be false. Another way of putting this is to say that a false theory can be empirically adequate but referentially empty. Nevertheless, such a theory remains causally connected to the world in the sense that we act on it, learn from it, and construct better theories. For a theory to function, then, it does not have to be true. If it turns out to have referential connections with the world, then that is an additional and rare bonus. So, to put the matter boldly: conceptions of leadership, whatever the specific constructs they contain, may turn out to be massively disconnected from the world and yet we can continue to talk as if they pointed to something real in terms of leaders 'turning the organization around'.

Take the example of leadership studies' best-known construct: initiating structure and consideration. These constructs were derived from an initial pool of 1790 items, were further reduced to 9 dimensions (Hemphill and Coons 1973), and eventually to the two remaining (Halpin and Winer 1973). These were considered the two smallest and most basic dimensions of leader behavior, and continue to be seen as such to this day. For brevity's sake, $S$ leadership was assumed to be in evidence, represented by a set of descriptive items, when a leader exhibited a requisite behavior such as, for example, 'defines his role and those of subordinates'; 'sets clear goals'; 'directs group activity through planning, communication, scheduling'. $C$ leadership, in contrast, was properly expressed in behaviors such as 'expresses appreciation for a job well done'; 'stresses the importance of high morale'; and 'is friendly' (Argyris 1979, p. 55). These sets of behaviors were seen as extensions of the relevant leadership conception and were believed to be unambiguously identifiable as per descriptive item.

Here it is important to remember the familiar point that what is taken to be as a relevant observable behavior is in part determined by our implicit, explicit or commonsense theory of leadership. What counts as an extension/reference depends upon the embedding framework or assumptions in which it is contained. But, most importantly, human behavior is always interpreted behavior. Put differently, it is human cognitive activity which holds the key to what is or is not counted as an extension, given an individual's 'cognitive economy' (Churchland 1979, p. 287) and the organizational (or other) context in which they find themselves. Put simply, what we believe to be a true reference regarding, say, democratic or C leadership, is a function of our theory of leadership which, in turn, is part of our changing, or ever developing, global theory of the world.

This does not mean to indicate that we have any kind of firm or secure grasp on leadership as a natural object. Reference is tied to our theories and assumptions and these can change, which means reference changes. So rather than beginning by wanting to prove or find evidence of presumed natural objects, or essences, such as leadership, we are better off by beginning from 'the ground up' by comparing the sentences of various leadership theories to see whether any of them answer to anything in the world. (Whether or not there are such things is a matter of our ontology and what evidence we can summon up by way of our epistemological resources.)

Returning to the point of specified behaviors as indicative of certain leadership styles, it becomes clear that these generic behaviors, supposed to be universally applicable and thus context-invariant, take on different meanings depending literally on in whose brain they appear. In other words, leadership constructs as conceptualized in empiricist science, which employs a hypothetico-deductive form of reasoning, fragment at the local level. For instance, the item 'expresses appreciation' does not map onto one specific interpretation of behavior but is open to a multitude of possible interpretations which may all be empirically adequate. This is where the category fragments because people just see things differently and in often widely discrepant ways, as demonstrated in studies Argyris (1979, p. 55) has conducted: 'In one study, 'Friendly and easily approachable foremen' (upon observation) turned out to be foremen 'who left the men alone and rarely pressured them' ... In another study 'friendly foremen' were those who took the initiative to discuss 'difficult issues' with the men'.

The category 'expresses appreciation' does not represent the local variance of interpreted behavior, rather it *abstracts* from the specificity of the local context. This is to be expected given that much of leadership research is based on the assumption of a functionalist framework. Recall Gibb's (1959, p. 917) most economical version: '... leadership is a function of personality and of the social situation, and of these two in interaction'; Hallinger and Heck's (1996, p. 6) more expansive yet compatible definition (which, however, simply identifies leadership with the principal's role, the formal office-holder), and an even more expansive description offered by Yukl (1973, p. 465). His 'linkage model' proposes to investigate the following:

> Leader behavior variables, intermediate variables, situational variables, subordinate preferences, criterion variables (i.e., satisfaction and productivity), and relevant leader traits ... [Add] Situational variables [such as] the organizational limiting conditions for participation ... the structural variables found to be associated with leader decision behavior ... the situational variables in Fiedler's model, the situational variables cluster-analyzed by Yukl ... and Woodward's (1965) system for classifying production technology.

Given such complexity, Yukl (1973, p. 465) advises that the predictive power of his model would be improved if one could 'identify which components of the behavior variables are the most important determinants of each intermediate variable'.

Stipulating functional relationships between such abstract concepts as 'situational, structural, and personal variables' begs the question since *any* empirical-material content fits the bill in that it can be subsumed under the abstractions. Since different situational and other factors obtain in different contexts, a functional explanation, which posits causal relations between individual relations, is not helpful. The point is not that functionalist explanations are merely abstract, the point is that they are vacuous, as well as pretentious (Evers and Lakomski 1991). This is another way of saying that the concept of leadership is causally massively disconnected from the world, and that no amount of sophisticated quantitative methodology makes any difference in its attempt to secure empirical results.

## Leadership Naturalized

The reason that the use of functionalist explanation and empiricist hypothetico-deductive theory obscures, abstracts from, and thus fails to capture the local and specific situations of leadership, resides in its limited view of human cognitive activity. All cognitive activity is identified with linguistic representation, and linguistic representation, in turn, is equated with knowledge, our scientific theories being the most austere examples.

Leaders' and followers' accounts of what they are doing duly recorded in surveys and questionnaires, are such sentential representations. But language as everyone knows, first has to be learnt, and learning itself is not primarily linguistic but is determined by electro-chemical brain activity. What we are able currently to represent in linguistic form, then, is only a relatively small part of all the cognitive activity which goes on in our brains. Much, or perhaps even most of our cognitive activity, such as the things we know how to do, cannot be represented linguistically because it is embedded in the fine-grained neuro-chemical circuitry of the brain, and is in part organized in neuronal patterns.

Given our preceding comments, it seems more productive to discontinue leadership studies, which appear to reduce quite readily to the study of effective administrative practice. If conceptions/theories of leadership fragment at the local level subject to organization members' individual interpretations, which, in turn, are a function of their shifting cognitive global economy, then theorizing about leadership/effective practice, at the deepest level, becomes a matter of explaining how the relevant neuronal patterns are activated which facilitate organizational (or any other) action. Several consequences would follow.

Leadership/effective administrative practice is a matter of local and highly specific factors, which cannot in principle, be universalized as postulated by empiricist theory. This means that large-scale prediction is not possible (Evers and Lakomski 1991). Whether or not there are general features in common between different organizational contexts would be a matter of empirical investigation, to be determined after the event by use of the coherentist criteria of our naturalism, rather than *a priori*, as was the case in hypothetico-deductive accounts. It may turn out to be the case that there is not one theory of leadership, but many, modular accounts.

A further consequence relates to organizational structure. If knowledge is not to be identified with the leader/or position, and presumed to flow from the top down, as traditionally assumed, then organizational functioning is much enhanced by gauging the knowledge of all organization members and structure the organization appropriately to feed it through all levels. Correction of possible error is thus to be emphasized since humans are fallible learners, and structure should prudently reflect human capacity.

These initial forays into a vast and complex field of study serve well to indicate the direction and magnitude of a challenging research agenda which promises spectacular and wide-ranging benefits: a naturalized account of leadership which explains effective organizational practice in schools and non-school organizations alike.

# References

Argyris C. (1979). How normal science methodology makes leadership research less applicable, in J.G. Hunt and L.L. Larson (eds.) *Crosscurrents in Leadership*. (Carbondale and Edwardsville: Southern Illinois University Press).

Bossert S.T., Dwyer D.C. and Lee G.V. (1982). The instructional management role of the principal, *Educational Administration Quarterly*, **18**(3), pp. 34–64.

Bowers D.G. and Seashore S.E. (1973). Predicting organizational effectiveness with a four-factor theory of leadership, in W.E. Scott and L.L. Cummings (eds.) *Readings in Organizational Behavior and Human Performance*. (Homewood, Ill: Richard D. Irwin).

Bridges E.M. (1982). Research on the school administrator: the state of the art, 1967–1980, *Educational Administration Quarterly*, **18**(3), pp. 12–33.

Churchland P.M. (1979). *Scientific Realism and the Plasticity of Mind*. (Cambridge, MA: Cambridge University Press).

Churchland P.M. (1993). *A Neurocomputational Perspective: The Nature of Mind and the Structure of Science*. (Cambridge, MA: MIT Press).

Evers C.W. (1995). Recent developments in educational administration, *Leading and Managing*, **1**(1), pp. 1–14.

Evers C.W. and Lakomski G. (1991). *Knowing Educational Administration: Contemporary Methodological Controversies in Educational Administration Research*. (Oxford: Pergamon).

Evers C.W. and Lakomski G. (1996). *Exploring Educational Administration: Coherentist Applications and Critical Debates*. (Oxford: Pergamon).

Fiedler F.E. (1967). *A Theory of Leadership Effectiveness*. (New York: McGraw-Hill).

Fiedler F.E. (1973). Validation and extension of the contingency model of leadership effectiveness: a review of empirical findings, in W. E. Scott and L. L. Cummings (eds.) *Readings in Organizational Behavior and Human Performance*. (Homewood, Ill: Richard D. Irwin).

Gibb C.A. (1959). Leadership, in G. Lindzey (ed.) *Handbook of Social Psychology*. (Reading, MA and London, UK: Addison-Wesley).

Hallinger P. and Heck R.H. (1996). Reassessing the principal's role in school effectiveness: a review of empirical research, 1980–1995, *Educational Administration Quarterly*, **32**(1), pp. 5–44.

Halpin A.W. and Winer B.J. (1973). A factorial study of the leader behavior description, in R.M. Stogdill and A.E. Coons (eds.) *Leader Behavior: Its Description and Measurement*. (Columbus, Ohio: College of Administrative Science, The Ohio State University).

Hemphill J.K. (1949). *Situational Factors in Leadership*. (Columbus, Ohio: Ohio State University Personnel Research Board).

Hemphill J.K. and Coons A.E. (1973). *Development of the Leader Behavior Description Questionnaire*. (Columbus, Ohio: College of Administrative Science, The Ohio State University).

Howe W. (1994). Leadership in educational administration, in T. Husén and T.N. Postlethwaite (eds.) *International Encyclopedia of Education*, 2nd Edition. (Oxford: Pergamon).

Hoy W.K. and Miskel C.G. (1996). *Educational Administration: Theory, Research and Practice*. (New York: McGraw-Hill, 5th Edition).

Korman A.K. (1966). Consideration, initiating structure, and organizational criteria — a review, *Personnel Psychology*, **19**, pp. 349–361.

Lawler E.F. and Porter L.W. (1967). The effect of performance on job satisfaction, *Industrial Relations*, **7**(1), pp. 20–28.

Perrow C. (1986). *Complex Organizations*. (New York: Random House).

Pitner N. (1988). The study of administrator effects and effectiveness, in N. Boyan (ed.) *Handbook of Research in Educational Administration*. (New York: Longman).

Stogdill R.M. (1948). Personal factors associated with leadership, *Journal of Psychology*, **25**, pp. 35–71.

Stogdill R.M. and Coons A.E. (eds.) (1957). *Leader Behavior: Its Description and Measurement*. (Columbus, Ohio: College of Administrative Science, The Ohio State University).

Yukl G. (1973). Toward a behavioral theory of leadership, in W.E. Scott and L.L. Cummings (eds.) *Readings in Organizational Behavior and Human Performance*. (Homewood, Ill: Richard D. Irwin).

# 5

# The Learning Organization

How an organization can best meet the challenges of an ever-changing social, physical, and economic environment and achieve its goals in the face of complexity and uncertainty remains a fundamental problem in the study of all social organizations, including those that deal with formal education. Given the fundamental nature of the problem, and the critical importance of solving it, it is not surprising that various disciplines and research programs attempt to throw light on what makes for effective organizational practice.

The purpose of this chapter is to examine one specific approach which addresses the issue of organizational practice, that of *organizational learning* and the *learning organization*. The learning organization is a concept which enjoys much prominence in current debates about restructuring in both public and private sector organizations, a debate mainly conducted in the broad fields of organizational and educational administration theory. It is advanced by its advocates as a possible solution to creating and maintaining effective organizational change. The argument presented here is that organizational learning is indeed a productive way to approach the problem of effective practice and ongoing organizational change, but that it can only fulfil its promise when it is augmented with a causal account of human cognition.

While organizational learning is probably the most ambitious attempt to describe and explain the effective functioning of a complex collectivity, even such respected advocates as Argyris and Schön (1996, esp. pp. 4–6) have little to say about what learning and human cognition consist in. Organizational learning remains a descriptive account which does not provide the required causal story to explain organizational experience, the practical suggestions of learning organization writers notwithstanding.

It is true to say, though, that in its emphasis on the complex and differentiated intra-organizational *relationships*, both structural and psychological, and on learning, the mainstream conception of organizational learning correctly highlights *organizational* functioning. In this respect, it represents a more comprehensive approach than that of leadership, for example, which remains preoccupied with the capabilities of individual leaders, albeit exercised in organizational contexts. It is worth emphasizing that the collaboration of individuals and groups to achieve organizational purposes — whether instigated by a leader or not — is fundamentally a *collective cognitive* task of immense complexity. It follows that any attempt to explain how this is accomplished, as indeed it is in everyday life, needs to be able to explain first what the individual human cognitive capacities are which make such feats possible. This goes for explanations of leadership as much as for explanations of how a whole organization can learn, and how it can be structured accordingly. Such emphasis is a first step in a naturalistic account of organizational learning and the learning organization. It does not underestimate the enormous complexities associated with sorting out issues of cultural context, organization as 'group mind', and specifying the sense in which organizations can know more than their individual members, all issues which have occupied social theoreticians at least since Mead wrote about them. We list them here to acknowledge that they are the subjects of much detailed future research for cognitive scientists in general and for our program in particular.

In keeping with the basic and modest requirement that whatever claims we make — about leaders or organizational learning, or whatever — have to be consistent with our ability to acquire and process knowledge in the first place, (the basics of *naturalistic coherentism* are found in Evers and Lakomski 1991; see further naturalistic extensions in Evers and Lakomski 1996) the neural net account of cognition and computation offers a much richer conception of knowledge representation than the standard symbolic account (see Evers and Lakomski 1996). It includes the kinds of non-symbolic, non-verbal activities which characterize practice, and it presents a reassessment of the role and function of symbol processing, such as language.

Furthermore, in extending the notion of individual to *socially distributed* or *cultural cognition*, connectionism is able to provide new tools to begin to explain how organizations can function as cognitive collectivities, and thus learn. The present discussion is a modest beginning which attempts to cast some light on how organization members learn; how organizations learn, and what some of the complexities are which surround the structuring of a learning organization so that it may capitalize on the collective cognitive capacities of all its members.

## Organizational Learning and the Learning Organization

The field of organizational learning is vast in scope, diffuse, and suffers from an absence of conceptual clarity and theoretical coherence, an observation which has led some researchers to warn that the concept of the learning organization has become 'a management Rorschach test' in which 'one sees whatever one wants to see' (Ulrich, Jick, and Von Glinow 1993, p. 57). (Both terms are used interchangeably for reasons which become clear in the following).

Organizational learning has its roots in the (North American) scientific management movement, and has been developed by eminent researchers such as Herbert Simon (1969), Peter Senge (1992) and Chris Argyris and Donald Schön (1978, 1996). The major problems which beset the field relate to analytical difficulties in the concept of organizational learning itself — whether to consider learning as new knowledge or insight (e.g. Argyris and Schön 1978, 1996; Hedberg 1981); as comprising new structures; new systems; mere actions, or a combination of all of the above (see Fiol and Lyles's 1985 overview who identify the preceding features; also Huber 1996). The terms learning, adaptation, change, innovation, and unlearning are all used in the attempt to account for organizational learning, and contribute little to clarifying this concept. Furthermore, what is also at issue is the question of the value, direction of, as well as threats to, organizational learning, and problems of implementation.

Although there have developed different definitions, conceptions, models, as well as different literatures, the field, following Argyris and Schön (1996), may be said to consist of two discernible camps: the practitioners who tend to talk about the 'learning organization' and the scholars who tend to be sceptical that there is any such thing as organizational learning. Practitioners and consultants tend to be prescriptive in approach, sometimes utopian in aspiration (see Senge 1992; Kofman and Senge 1995), and mainly concerned with the 'how to' of structuring learning organizations. The researchers, on the other hand, tend to be concerned with the very meaning of organizational learning, and attempt to explain what it might consist in. Nevertheless, although the audiences of the two domains differ, as do their respective literatures, there is broad agreement on the main ingredients of the ideal of 'the learning organization'. In his lucid overview, Dodgson (1993, p. 377) provides a broad definition of organizational learning which contains the features considered central by most researchers:

> Learning ... relates to firms, and encompasses both processes and outcomes. It can be described as the ways firms build, supplement and organize knowledge and routines around their activities and within their cultures, and adapt and develop organizational efficiency by improving the use of the broad skills of their workforces.

Argyris and Schön's (1996, p. 180) definition, while also broad is more specific:

> [Organizational learning] includes notions of organizational adaptability, flexibility, avoidance of stability traps, propensity to experiment, readiness to rethink means and ends, inquiry-orientation, realization of human potential for learning in the service of organizational purposes, and creation of organizational settings as contexts for human development. (See also McGill and Slocum, Jr. 1993).

Speaking from within the behavioral studies of organization tradition, Levitt and March (1988, p. 319) are more parsimonious: 'Organizational learning is ... routine-based, history-dependent, and target-oriented. Organizations are seen as learning by encoding inferences from history into routines that guide behavior.' We might just simply say that they learn from experience.

Assumptions generally but not unproblematically shared in the field of organizational learning are (1) learning has positive consequences, and includes learning from error; (2) the organization learns as a whole: the learning of individuals is not identical with the latter, a distinction sometimes blurred, and (3) learning occurs throughout the organization, happens at different speeds and comprises both lower-level and higher-level learning. The definition of a learning organization is one which 'purposefully construct[s] structures and strategies so as to enhance and maximize organizational learning' (Dodgson 1993, p. 377). Importantly, and this is a point taken up again later, as Levitt and March note, organizations learn from direct experience as well as the experience of others; construct interpretive frameworks for that experience; encode, store, and retrieve experience independent of individual agents or the passage of time, and also discard knowledge which is no longer useful, that is, they *unlearn*, a process which seems particularly problematic (see Hedberg 1981).

As for differentiating the type and level of learning in an organization, there is a general distinction between (a) learning which merely 'adjusts parameters in a fixed organizational structure' (lower-level learning), and (b) the kind which 'redefines the rules and changes the norms, values and world views' (higher-level learning) (Fiol and Lyles 1985, p. 806). Lower-level learning (which can happen at any organizational level) is based on routine, affects only part of the organization, tends to happen in organizational contexts which are well understood, and results in a particular behavioral outcome or level of performance. This kind of learning has come to be best known in Argyris and Schön's (1978, 1996) terminology as 'single-loop learning'. Initiated by an organizational inquiry in response to the detection of a mismatch of expectation to outcome, single-loop learning is 'instrumental learning that changes strategies of action or assumptions underlying strategies in ways that leave the values of a theory of action unchanged' (Argyris and Schön 1996, p. 20). The resultant modification ensures that organizational performance returns to the range preset by existing norms and values. In contrast, higher-level, or 'double-loop learning' is far-reaching in its consequences since it questions and adjusts the organization's values and norms, and ways of doing things. It 'results in a change in the values of theory-in-use, as well as in its strategies and assumptions' (Argyris and Schön 1996, p. 21). In addition to the two kinds/levels of learning Argyris and Schön discuss, they define a third which is a variation of double-loop learning: *deuterolearning* which involves 'learning how to learn' and by means of which '*members of an organization may discover and modify the learning system that conditions prevailing patterns of organizational inquiry*' (Argyris and Schön 1996, p. 29. Italics in original).

As previously noted, one of the most important consequences of double-loop (and deutero) learning may result in the organization's willingness to rid itself of behaviors which have outlived their usefulness, although this by itself is also an issue fraught with analytical difficulties which go beyond the scope of this chapter (but see Hedberg 1981, for a good discussion of this issue).

In addition to the types and levels of learning identified in organizations, there is a further important distinction to note: Argyris and Schön's conception of the organization's *theory-of-action*, the espoused theory, and its *theory-in-use*, the 'lived' theory. The former indicates what an organization, for example, is officially about, as espoused in its organizational chart, policy statements and job descriptions. This implies the organization's strategies for action, its norms and assumptions. This 'official' theory is opposed to the often tacit theory-in-use, the way members actually go about doing their work which may differ significantly from what they *say* they do. In that sense it is the more important because it accounts for the organization's identity over time. In the authors' view,

> A theory-in-use is not a 'given' ... In the case of organizations, a theory-in-use must be constructed from observation of the patterns of interactive behavior produced by individual members of the organization, insofar as their behavior is governed by formal or informal rules for collective decision, delegation, and membership (Argyris and Schön 1996, p. 14).

Argyris and Schön's tacit theory-in-use captures the element of practice which is often circumscribed as 'knowing more than we can tell' (Lakomski 1997).

Finally, it is important to identify some of the factors which are said to be problematic in an organization's learning. One observation is uncontested in the field: organizations are relatively good at single-loop learning but poor at double-loop learning. In fact, in Argyris and Schön's studies there were no organizations which demonstrated either form of higher learning. Suffice it to note here their extensive discussions on the various defence mechanisms both individuals and groups use in terms of primary (at the individual level) and secondary (at the group level) inhibitory loops (Argyris 1990). These loops tend to create learning systems which prevent organizations from double-loop learning. Many other inhibitors are detailed by Levitt and March (1988, p. 322), including 'competency traps', situations 'when favorable performance with an inferior procedure leads an organization to accumulate more experience with it, thus keeping experience with a superior procedure inadequate to make it rewarding to use'. Sticking with old procedures in the face of newer and better ones is, from a social and evolutionary point of view, not a wise thing to do since it leads an organization to persist with procedures or technologies which are no longer optimal. A second difficulty resides in the fact that members need to *interpret,* that is, draw inferences from experience, a fact which raises many problems (Kahneman *et al.* 1982; Nisbett and Wilson 1977). Finally, for present purposes, there is 'superstitious learning', which denotes an instance of successful subjective learning but mis-specifies 'the connections between actions and outcomes'.

While the above are only a few of the many problems identified in the literature which prevent an organization from higher level learning, there is — in the face of such complexity — nevertheless agreement among researchers that organizations do, in fact, learn while they remain agnostic on the issue of how learning is accomplished. Although Argyris and Schön (1996, pp. 42–43) make reference to the fact that practitioners' causal inquiry consists in 'Their situation-specific inferences of design, efficient, or pattern causality', that is, constructing causal stories which, as prototypes, are carried over from old to new situations (reflective transfer), how such pattern constructing can be accomplished is not explained. Here we can provide some answers which contribute to an understanding of the causal processes involved, both in individual and organizational learning. To this issue we turn in the next section by way of raising some more problems in the conception of organizational learning, and considering an alternative, non-cognitive view.

## Cognition, Culture, and the Distribution of Cognitive Labor

It has, of course, not gone unnoticed in the organizational learning literature that knowledge of how individuals acquire and process knowledge is important in any account of organizational learning. In fact, the 'cognitive perspective' in organizational learning (Cook and Yanow 1996) constitutes the main thrust of the field, and March and Olsen, Argyris and Schön, and Simon are its prominent representatives. The latter, whose cognitive views have shaped a number of research enterprises, also contributes to defining organizational learning. In his view

> All learning takes place inside individual human heads; an organization learns in only two ways: (a) by the learning of its members, or (b) by ingesting new members who have knowledge the organization didn't previously have (Simon 1996, p. 176).

One of the field's most central problems is how to conceive of the relationship between learning by individuals *in* organizations and of learning *by* organizations as separate entities. Hedberg (1981, p. 6), for example, expresses one prominent view of how this relationship ought to be considered. He observes, 'there are many similarities between human brains and organizations in their roles as information-processing systems. Insights into how human brains process and store information can suggest important processes in organizations' learning.' The view expressed is not uncontested in the mainstream debates of organizational learning, as we will see in the following. More specifically, the treatment of organizations as if they were brains in the mainstream literature hides a number of conceptual and empirical problems which make it difficult to provide a satisfactory account of individual learning and its relationship to organizational learning, as well as the nature of *organizational* learning itself.

For our purposes it is helpful to lay out what these problems are, and then to show how they can be suitably reframed in our naturalistic account of cognition, including that of organizational cognition. A useful way of doing this is by way of discussing an alternative view to the cognitive perspective, in particular that offered by Cook and Yanow (1996) who argue for a *cultural perspective of organizational learning* as a complement, not a replacement, to the cognitive view. We will have occasion in the final sections of this chapter to provide some answers to the problems raised in both mainstream and cultural perspective views.

Beginning with the troublesome relationship of individual learning *in* organizations and learning *by* organizations, there are at least two ways in which this relationship has been conceived in the cognitive perspective characteristic of the mainstream organizational learning literature which still dominates the field. On one hand, individual learning is considered to be a particular kind of learning undergone in organizations by key personnel who are then instrumental in effecting change; on the other, organizations themselves are believed to be imbued with learning capacities akin to individual capacities, and it is these which constitute organizational learning. In the second case, organizations are presumed to be like individuals, the view held by Hedberg amongst others. According to Cook and Yanow (1996) though, both approaches begin from the same background assumption: an understanding of how *individuals* learn, an assumption they believe to be mistaken as a starting point. There is, in their view, a kind of learning which goes beyond that of both approaches described. While there is nothing inherently wrong in applying models of individual learning to organizations, they query whether such individual learning adequately describes learning by a whole organization, either empirically or conceptually. They warn (Cook and Yanow 1996, p. 435):

> … because it is not obvious … that organizations are cognitive entities, in drawing on individual cognition as a way of understanding organizational phenomena, we must take care not to lose sense of the 'as if' quality of the metaphor, forgetting that organizations and individuals are not the same sort of entities.

This worry is also echoed by Simon in that the potential reification of organizations ought to be considered. For Cook and Yanow the problem is avoided when organizations are primarily considered as *cultural* entities, not as cognitive ones (see also Schein 1985). In addition to sidestepping the issue of reification, such cultural emphasis is also more productive because there is, in their view, no commonly accepted model of individual cognition yet. In addition, the cultural view also addresses a third problem inherent in the cognitive perspective: the assumption that individual learning is the same as organizational learning. As Cook and Yanow (1996) see it, this is far from obvious and individuals and organizations cannot be assumed *a priori* either to learn in the same way, or that the outcomes of their learning would be identical, even if one presumed similar cognitive capacities. A fourth problem raised is the twinning of organizational learning with organizational change and increased effectiveness, and, conversely, the assumption that no learning has taken place in the absence of observable change. In either case though, as is readily seen, change may have happened without any outward manifestation; or where effectiveness is concerned, we may have learnt destructive rather than constructive habits, false concepts, or inappropriate skills (also Huber 1996, p. 126). Cook and Yanow agree that organizations can indeed learn but that the question of organizational learning 'is not an epistemological one about cognitive capacities, but an empirical one about organizational actions — (Cook and Yanow 1996, p. 431). Equating organizational learning with observable change is unnecessarily limiting, especially in light of the fact that there are other notions of learning. The kind of (organizational) learning they have in mind is indicated in their example 'The Celtics know how to play basketball' (Cook and Yanow 1996, p. 438). This collective activity both encapsulates organizational knowing and learning which can only be done by a group. The same description applies to the playing of a Schubert symphony by the Melbourne Symphony Orchestra, or the making of world-class flutes, an example provided by the authors.

The essential features in all these examples is that organizational learning consists in the 'capacity of an organization to learn how to do what it does, where what it learns is possessed not by individual members of the organization but by the aggregate itself. ... when a group acquires the know-how associated with its ability to carry out its collective activities, that constitutes organizational learning' (Cook and Yanow 1996, p. 438). Culture in this context is a 'set of values, beliefs, and feelings, together with the artifacts of their expression and transmission (such as myths, symbols, metaphors, rituals), that are created, inherited, shared, and transmitted within the one group of people and that, in part, distinguish that group from others' (Cook and Yanow 1996, pp. 439–440).

As they make clear in their examples of flute making, the creation of a Powell Flute (Cook and Yanow 1996, pp. 440–448), considered one of the finest in the world, is not the accomplishment of the individual craftsman, no matter how skilled, but is a collective and hence organizational feat. No one individual combines all the knowledge and skills required to create a flute, but together, the task can be accomplished with excellence, and over a long period of time, where individuals leave, as well as join, the workshop. In this sense, organizational learning differs from individual learning, and learning can legitimately be described as having been done by the group. The authors do not consider this a cognitive activity because organizations do not possess the requisite wherewithal for cognition, such as having actual bodies, organs and brains. What can be noted, however, and thus becomes a matter of empirical observation, is the expression in *group action* of intersubjective meanings as expressed in and through human artifacts including objects, language, symbols, ceremonies, and acts. The group as a collectivity is thus the primary focus of analysis and observation. The relevant question now becomes 'What is the nature of learning when it is done by organizations?' (Cook and Yanow 1996, p. 440). In this way the emphasis has shifted from attention to the individual to that of a collectivity, and this is a considerable shift in focus between the cognitive and the cultural perspectives. The definition of organizational learning thus arrived at is as '*the acquiring, sustaining, or changing of intersubjective meanings through the artifactual vehicles of their expression and transmission and the collective actions of the group*' (Cook and Yanow 1996, p. 449. Emphases in original). In addition to this shift in emphasis, what makes the cultural perspective richer is the incorporation of the tacit dimension of practice and communication which contrasts sharply with the cognitive view 'which typically requires that those things essential to organizational learning be made explicit, so that they can be communicated' (Cook and Yanow 1996, p. 449).

As is evident from this characterization of the cultural perspective, organizational learning is to be explained both through cognition, as understood in the mainstream literature, as well as through the cultural aspects the authors discuss, where the latter are seen as non-cognitive. We are now in a position to ask how cognition and culture can be appraised from our naturalistic perspective aimed at explaining organizational learning and practice.

Providing answers to both issues, and assessing the valuable contribution of the cultural perspective in particular is made possible by recent developments in connectionism, most centrally by the idea of *distributed cognition*. But before we discuss *distributed* cognition, and since this is not the place to rehearse the fundamentals of the neural net account as explained in the new cognitive science, let us just note that we have an answer to Cook and Yanow's assertion that there is no suitable account of individual cognition and learning. It is of course the new cognitive science which makes it possible to unite the cognitive with the cultural components of learning and thus break down a dichotomy which still characterizes the field of organizational learning, as seen in Cook and Yanow's otherwise very perceptive work. As will be seen in the following, the extension of cognition from individually owned to culturally and socially embedded capacity facilitates an understanding of cognition as cultural and culture as cognitive; such understanding has profound consequences for the conception of organizational learning and for organizational structure. Not all questions raised in the organizational learning literature can yet be answered, but it is possible to see the contours of a naturalistic conception of organizational learning and what form the learning organization might take. The most important recent development is the conception of *distributed cognition*.

*Distributed cognition* is able to cast new light on what has traditionally been described as an organization's cognitive systems, memories, mental maps, and symbols, and begins to explain their function in the cognitive economy of social life. It is thus possible to provide some fresh suggestions on the relationships between individual group, and organizational levels which 'has been poorly described in the literature', and provide some research results on the interaction between learning individuals and learning organizations which 'is badly needed' (Hedberg 1981, p. 6). As a consequence, we also suggest further reflections on organizational design.

To begin with, one of the most familiar and accepted concepts in organizational theory and in everyday life is that of the division of labour. It implies a dividing up of tasks, a design for the performance of work and communication paths, all encapsulated in its structure. It also implies a *division of cognitive labour*. This is an idea which, surprisingly, has escaped the attention of most organization theorists to date. (An early exception is March and Simon's 1958 book *Organizations* where the conception of cognition however was equated with symbol processing).

We know now that human brains do not think the way a computer does (see Chapter 3, this volume). Computer thinking is rule-governed symbol manipulation whereas human cognition includes the ability to manipulate symbols, e.g. speak, write and number-crunch, but is not exhausted by this ability. Rather, given the vast interconnectivity of the human brain in terms of the interrelationships of whole systems of neural nets, we are able to process information and carry out computation tasks in parallel fashion (parallel distributed processing). The brain's ability to activate prototypes in response to whatever external input it receives makes us good at pattern recognition and completion, that is, at making plans, identifying faces (Churchland 1995, 1996), recalling and recognizing relevant information, and understanding speech (Tienson 1990, p. 382). Without going into the details of neural net architecture here (see Evers and Lakomski 1996, especially Chapter 8), the pattern of activation (prototype) which is your daughter's face, for example, would consist in the strength of the connections between neuron-like processors, or their weight. Your daughter's face is thus internally *represented* in the connection strengths obtaining in a particular neuronal configuration. In neural nets many units or processors are involved in enacting representation, hence we speak of *distributed* representations. Information in connectionist systems is actively represented as a pattern of activation and is not stored as data structures. It is stored in its weights, and this means that it is created and re-created as the situation requires. Learning on the connectionist account consists in the changing of the weights. In the case of understanding language, for example, it is not the case that we store linguistic information about each individual sentence. Rather, 'In understanding a sentence, we create linguistic information about it on the spot, in response to present stimuli. Except in the case of a few cliches, it does not matter whether we have heard the sentence before or not.' (Tienson 1990, p. 393). Representations in connectionist systems do not have a syntactic structure. The reasonable conclusion to be drawn — from this very simplified account — is that human cognition and reason includes linguistic-symbolic representation, but that this kind of representation is only a minor part of cognition which is softer and more associative in nature, a point central to all discussions of administrative or other practices in this volume.

Another assumption which went along with the notion that humans think like a computer was the idea that just as a computer program belongs inside a computer, so cognition belongs inside an individual's skull. This also had the consequence that culture, context and history were considered to be 'outside' of human cognition and could be studied in isolation from it (Hutchins 1996, Chapter 9). But this individual ownership model of human cognition has also come under attack from research programs other than connectionism such as 'situated action' or 'situated cognition' which we discussed in detail in Chapter 3. The cultural perspective offered by Cook and Yanow (1996) is yet another alternative which challenges the cognitive view from a related but not identical perspective.

What unites these approaches in all their diversity is the view that although human knowledge is constructed through direct experience, humans also build their knowledge through what others tell them, in written texts, pictures, and gestures. It is thus more appropriate to describe knowledge 'as distributed across several individuals whose interactions determine decisions, judgments and problem solutions' (Resnick 1991, p. 3). There are as yet few empirical studies which describe how people reason and learn in concrete settings (e.g. Lave 1991; Scribner 1984;

Rogoff and Lave 1984; Levine and Moreland 1991; Hastie and Pennington 1991). They are important because they demonstrate how agents 'learn by doing' in interaction with each other and the specific environmental circumstances in which they find themselves (a way of conceiving of learning in the naturalistic tradition of Dewey). Importantly, in detailing the minutiae of practice, the point made by these researchers is that knowledge is 'constructed' in the doing, not in the being told. This, of course, is only possible, as noted above, since knowledge acquisition and processing is not primarily a symbol manipulation exercise. It is this broader notion of cognition which makes it possible to include the making of flutes as described by Cook and Yanow as a cognitive enterprise without needing to give up the cultural and group aspect of the task. It is both.

The only full-scale *connectionist* anthropological study to date which exemplifies cultural cognition is Hutchins' (1996) marvellous work on ship navigation to which we frequently refer. He argues that culture is a human cognitive process that takes place both inside and outside the minds of people. Delimiting cognitive systems in terms of the inside/outside boundary, as has been sanctioned by classical cognitive science, and considering all intelligence as being located on the inside of that boundary, has the major disadvantage of overattribution. As a consequence, we might erroneously attribute the right properties to the wrong system, or even invent the wrong properties and attribute them to the wrong system (Hutchins 1996, p. 355). Studying 'cognition in the wild' (the title of his book) is the best way to understand the mutually interacting cognitive properties of the sociocultural system and the cognitive properties of humans who manipulate these elements (Hutchins 1996, p. 362).

Since cognition is 'spread' the way it is, it naturally interacts with, creates, and maintains both symbolic and social-institutional structures. A simpler way of putting this is to say that the mind/brain offloads some important problem-solving tasks onto external structures since many computational tasks are too complex for one individual mind (Clark 1997). The collective computations of external structures in turn exhibit their own special dynamics and properties which then react back onto, and shape, individual cognition in reciprocal interaction. Organizations are a natural embodiment of this point. Since we are creatures of finite cognitive capabilities, the dissipation of reason is a necessity of our nature without which we would not have been able to acquire the kind of advanced cognition which enables us to construct and manage organizations in the first place. External structures are extensions of our minds, and organizations and institutions provide the external resources in which we conduct our affairs, given their policies, practices and norms. Linguistic artifacts, including ever important administrative rules, regulations and policies, thus play the part of other external structures. They help us reshape some cognitive tasks so that we can cope with them.

If our minds indeed 'leak' into the world (Clark's colourful phrase), and if the external structures of the world are augmentations of our individual cognition, what consequences can be drawn from such a wide conception of cognition for organizations, organizational learning and design?

## How to Conceptualize a Learning Organization as a System of Socially Distributed Cognition

Our approach in this chapter proceeded on the modest naturalistic assumption that whatever claims we make — on behalf of organizational learning, or whatever — have to be compatible with our natural capacities for acquiring and processing knowledge. The view of the human mind and of knowledge in the empiricist leadership tradition, for example, was too narrow to shed light on how leaders manage to do what they do; it thus explained little (see Chapter 4, this volume); and to the extent that the organizational learning literature remains agnostic on how we learn, the idea of organizational learning also falls short in regard to its central explanatory feature. So what are some of the immediate conclusions to be drawn from our knowledge of human cognition as pattern recognition and completion for organizational learning and organizational structure?

*Naturalistic coherentism* allows us to consider organizational theories of action, described by Argyris and Schön, as external, symbolic-linguistic, augmentations of practitioners' cognition. They have thus a more active part to play than traditionally assumed, although what the precise nature of the interplay between, say, policies, routines and individual cognition is, remains a huge research task for the future. *Theories-in-use,* or practitioners' practical knowledge, is now explicable in terms of the (non-linguistic) prototype activation patterns enacted by the parallel distributed brain in response to both external as well as internal stimuli. While such knowledge is not represented in symbolic form, it nevertheless is represented; except its form of representation is in the connection weights of an activation pattern. Thus we have a much richer account of the representation of human reason and intelligence which includes non-verbal behavior and practices which hitherto remained mysterious but are what effective practice consists in. On this account it becomes comprehensible how the organizational know-how, represented in the individual flute makers, comes to be when they judge a Powell flute as finished when it 'has the right look or right feel' (Cook and Yanow 1996, p. 444).

We also have a richer account of learning which may be said to consist in the changing of the weights of enacted patterns of activation in interaction with whatever external features, linguistic or otherwise, make up the specific learning situation, including those in organizational contexts. The structures and processes of our places of work are quite literally extensions of our minds, and reciprocally shape what we know, and vice versa. In this manner, a better defensible account of organizational learning may be constructed.

It is now possible to expand on an answer to the question of organizational learning which, while raised earlier in this discussion, is formulated by Argyris and Schön in this perceptive manner: 'What is an organization that it may learn?'

An organization, on the above account, is a system of socially distributed cognition. Without falling back into the assumptions of the cognitive perspective with its associated problems which have been answered by connectionism and our naturalism, we may consider an organization as being like a brain. It is a living example of distributed information processing and problem solving which embodies a variety of strategies of decomposition and coordination (on the advantages and disadvantages of distribution in naturally occurring systems see Chandrasekaran 1981; Fox 1981). In order to understand, first of all, how it distributes its cognitive labor, it is important to understand the interplay between individual and group-level cognition, a classical problem in organizational theory and organizational learning. (For an interesting discussion of the problem of group mind see Weick and Roberts 1996). Secondly, we need to understand what difference the *social organization* of distributed cognition makes in organizational contexts. These are deep and uncharted waters, and we will only be able to make some sketchy comments to indicate the direction of future research.

On a neural net account, an individual's cognition may be modelled by whole networks or assemblies of networks. Systems of socially distributed cognition such as groups can be modelled by communities of networks (Hutchins 1991, p. 293). The focus for the study of organizational learning/functioning then becomes: how do the networks interact with each other, and do they interact differently, given different patterns of social organization? The question matters greatly because if it turns out that social organization makes a difference in group problem-solving, then this has ramifications for designing appropriate structure to maximize group cognition. To put the matter another way: do groups, apart from the individuals who comprise them, have cognitive properties of their own, a question which is the central worry of the traditional organizational learning literature. Hutchins examined this issue, and came to some interesting results. He put the research issue like this:

> ... if groups can have cognitive properties that are significantly different from those of the individuals in the group, then differences in the cognitive accomplishments of any two groups might depend entirely on differences in the social organization of distributed cognition and not at all on differences in the cognitive properties of individuals in the two groups (Hutchins 1991, p. 285; see also Hutchins 1996, Chapter 5).

Since increased or better communication between groups (as in decentralized structures like the learning organization) is often advocated as enhancing organizational efficiency, Hutchins set out to provide evidence for what we know 'from experience': more communication is not always better, and more heads don't always know more than one (Levine and Moreland 1991).

In a simulation exercise Hutchins (1991, 1996) examined this complex interplay by studying the ways in which various patterns of communication shape the problem-solving behavior of small groups. In particular, he set out to discover how *confirmation bias*, a property of individual cognition, affects the behavior of groups in terms of finding an optimal interpretation/solution for a problem encountered. Confirmation bias is the human propensity to hold on to an already formed interpretation and 'to discount, ignore, or reinterpret evidence' (Hutchins 1991, p. 286) which might contradict it (also Clark 1997, pp. 187–188). Since confirmation bias is an individual property involving a trade-off between a sub-optimal or no interpretation, how could a collection of individuals be organized so that it might reach the best possible interpretation and neutralize individual bias?

In his simulation exercise, each individual was considered as a small neural net with a limited number of linked processing units, and each unit was coded for a specific feature in the environment. Mutually supportive features were linked by excitatory links, and mutually inconsistent features were connected with inhibitory features. The network which is coded 'is a dog', for example, would edit out 'meows', but would activate an excitatory link to 'barks' (Clark 1997, p. 187). These kinds of networks are called *constraint satisfaction networks*, and provide a rough model of individual interpretation formation (Hutchins 1991, p. 289). Considering a community of such constraint satisfaction networks where each network has a different initial activation level and different access to environmental data, Hutchins found that the way the inter-network communication was structured had a marked effect on the communities' collective problem-solving. In the case where communication was shared evenly between all the networks, that is, where networks were allowed to influence the activity of the others, the overall system showed a remarkable degree of confirmation bias. This is so because the density of the communication patterns force the rapid move toward a stable pattern across all units; that is, to settle on a shared interpretation of the data. External input is ignored in favour of satisfying the internal constraint of finding stable activation patterns across all units (see Hutchins 1991, pp. 297–299). In other words, the system behaves as one large net. However, where the level of early communication is initially restricted, individual nets have the opportunity to balance their predispositions against the external input data. If communication is subsequently enabled, confirmation bias reduces. The group, unlike its individual members, is better able to find the best solution.

The results this simple simulation has yielded have interesting ramifications for group decision-making in organizations. With regard to jury decision-making, for example, it suggests that the more information the jury gets from the beginning, the less it is able to cast about for the best decision. This means that its collective advantage is lost proportional to the amount and timing of information it receives. The same consideration applies to decision-making in schools, and especially in those which have decentralized structural arrangements in the belief that such arrangements implicitly improve their decision-making. So here we need to insert a cautionary note in the face of advocating decentralized, or 'heterarchical' structures (Fox 1981) in view of the possible trade-offs which have to accompany it.

Despite this, what seems clear is that the idea of structural hierarchy (which was implicit in the hypothetico-deductive account of leadership) is unhelpful. As we have argued elsewhere (Evers and Lakomski 1996, Chapters 5 and 6), the different view of organizational structure which emerges is *decentralized*, whatever else it might turn out to be in terms of possible trade-offs. Here our argument seems to mesh quite happily with the learning organization's most characteristic feature: a decentralized structure which, when combined with appropriate feedback loops, facilitates organizational learning. This is just what one would expect from a connectionist point of view, and we can now see the reasons why.

So, what has changed, and what has remained? In one sense, the present discussion can be seen as a redescription of our commonsense understandings of organizational learning with the help of current neural network accounts of learning and cognition. On the surface, as it were, much of the practical proposals advocated by learning organization writers remain in place as sensible. What has changed is the account of how they are made to work. A naturalistic account of individual and group cognition, and the social-cultural organization of the distribution of cognitive labor remain large topics for future work. Nevertheless, what we hope we have shown in this chapter specifically, based more generally on our previous work, is that Cook and Yanow's assertion that the question of organizational learning is not an epistemological but an empirical issue about organizational actions, can be rebutted. Our naturalism is able to account for both in that there are no principled divisions between epistemology and empirical reality. Constraints on theorizing which pertain to the classical view of organizational learning, as well as to the cultural perspective alternative, are not part of naturalistic coherentism which is 'constrained' by the best current account of human cognition in its symbolic and particularly non-symbolic features, and recently extended into social and cultural contexts. We thus believe that our account presents a much richer and better defensible explanation of both individual and organizational learning. As for further developments regarding organizational structure, Fox (1981, p. 70) rightly comments 'Distributed systems are difficult to design', and what an optimal design for a structure such as school may look like we can but guess at this point in time. Nevertheless, with the naturalistic exploration of 'cognition in the wild' well and truly under way, in this and related areas of human practice, the task is at least charted in the right direction.

# References

Argyris C. (1990). *Overcoming Organizational Defenses*. (Needham Heights, MA: Allyn and Bacon).

Argyris C. and Schön D.A. (1978). *Organizational Learning*. (Reading, MA: Addison-Wesley).

Argyris C. and Schön D.A. (1996). *Organizational Learning II*. (Reading, MA: Addison-Wesley).

Chandrasekaran B. (1981). Natural and social system metaphors for distributed problem-solving: Introduction to the issue, *IEEE Transactions on Systems, Man, and Cybernetics*, Volume SMC-**11**, 1, pp. 1–5.

Churchland P.M. (1995). *The Engine of Reason, The Seat of the Soul*. (Cambridge, MA: MIT Press).

Churchland P.M. (1996). The neural representation of the social world, in L. May, M. Friedman, and A. Clark (eds.) *Mind and Morals*. (Cambridge, MA: MIT Press).

Clark A. (1997). *Being There: Putting Brain, Body, and World Together Again*. (Cambridge, MA: MIT Press).

Cook S.D.N. and Yanow D. (1996). Culture and organizational learning, in M.D. Cohen and L.S. Sproull (eds.) *Organizational Learning*. (Thousand Oaks, CA: Sage).

Dodgson M. (1993). Organizational learning: A review of some literatures, *Organization Studies*, **14**(3), pp. 375–394.

Evers C.W. and Lakomski G. (1991). *Knowing Educational Administration*. (Oxford: Elsevier).

Evers C.W. and Lakomski G. (1996). *Exploring Educational Administration*. (Oxford: Elsevier).

Fiol C.M. and Lyles M.A. (1985). Organizational learning. *Academy of Management Review*, **10**(4), pp. 803–813.

Fox M.S. (1981). An organizational view of distributed systems, *IEEE Transactions on Systems, Man, and Cybernetics*, Volume SMC-**11**, 1, pp. 70–80.

Huber G.P. (1996). Organizational learning, in M. D. Cohen and L. S. Sproull (eds.) *Organizational Learning*. (Thousand Oaks, CA: Sage).

Hastie R. and Pennington N. (1991). Cognitive and social processes in decision making, in L. B. Resnick, J. L. Levine and S. D. Teasley (eds.) *Perspectives on Socially Shared Cognition*. (Washington, DC: American Psychological Association).

Hedberg B. (1981). How organizations learn and unlearn, in P. C. Nystrom and W. H. Starbuck (eds.) *Handbook of Organizational Design*. (Oxford: Oxford University Press).

Hutchins E. (1991). The social organization of distributed cognition, in L. B. Resnick, J.M. Levine, and S.D. Teasley (eds.) *Perspectives on Socially Shared Cognition*. (Washington, DC: American Psychological Association).

Hutchins E. (1996). *Cognition in the Wild*. (Cambridge, MA: MIT Press).

Kahnemann D., Slovic P. and Tversky A. (1982). (eds.) *Judgment Under Uncertainty: Heuristics and Biases*. (Cambridge: Cambridge University Press).

Kofman F. and Senge P.M. (1993). Communities of commitment: The heart of learning organizations, *Organizational Dynamics*, **22**(2), pp. 5–9.

Lakomski G. (1997). Tacit knowledge in teacher education, in *Volume 2*: *Logical Empiricism in Educational Discourse (Advanced Volume) D. N. Aspin* (ed.) (Durban, SA.: Butterworths).

Lave J. (1991). Situating learning in communities of practice, in L.B. Resnick, J.L. Levine and S.D. Teasley (eds.) *Perspectives on Socially Shared Cognition*. (Washington, DC: American Psychological Association).

Levine J.M. and Moreland R.L. (1991). Culture and socialization in work groups, in L.B. Resnick, J. M. Levine and S. D. Teasley (eds.) *Perspectives on Socially Shared Cognition*. (Washington, DC: American Psychological Association).

Levitt B. and March J.G. (1988). Organizational learning, *Annual Review of Sociology*, **14**, pp. 319–340.

March J. and Simon H.A. (1958). *Organizations*. (New York: Wiley).

McGill M.E. and Slocum Jr., J.W. (1993). Unlearning the organization, *Organizational Dynamics*, **Autumn**, pp. 67–79.

Resnick L.B. (1991). Shared cognition: thinking as social practice, in L.B. Resnick, J.M. Levine and S.D. Teasley (eds.) *Perspectives on Socially Shared Cognition*. (Washington, DC: American Psychological Association).

Resnick L.B., Levine J.M., and Teasley S.D. (1991). (eds.). *Perspectives on Socially Shared Cognition*. (Washington, DC: American Psychological Association).

Rogoff B. and Lave J. (eds.) (1984). *Everyday Cognition*: *Its Development in Social Context*. (Cambridge, MA: Harvard University Press).

Schein E. (1985). *Organizational Culture and Leadership*. (San Francisco: Jossey-Bass).

Scribner S. (1984). Studying working intelligence, in B. Rogoff and J. Lave (eds.) *Everyday Cognition*: *Its Development in Social Context*. (Cambridge, MA: Harvard University Press).

Senge P. (1992). *The Fifth Discipline*. (New York: Doubleday).

Simon H. (1996). Bounded rationality and organizational learning, in M. D. Cohen and L. S. Sproull (eds.) *Organizational Learning*. (Thousand Oaks, CA: Sage).

Tienson J.L. (1990). An introduction to connectionism, in J. L. Garfield (ed.) Foundations *of Cognitive Science*: *The Essential Readings*. (New York: Paragon House).

Ulrich D., Jick T. and Von Glinow M.A. (1993). High-impact learning: building and diffusing learning capability, *Organizational Dynamics*, **Autumn**, pp. 52–67.

Weick K.E. and Roberts K.H. (1996). Collective mind in organizations, in M.D. Cohen and L.S. Sproull (eds.) *Organizational Learning*. (Thousand Oaks, CA: Sage).

# 6

# Natural Decision-Making

The field of decision-making, even when restricted to the domain of educational management, is vast and complex. Our purposes, on the other hand, are limited and modest. In what follows, we propose to focus on some of the issues concerning educational decision-making in *naturalistic* settings. (For a similar approach, see Klein *et al.* 1993, and Zsambok and Klein 1997.) That is, we are concerned with decision-making 'in the wild' or, to switch metaphors, 'on the hoof' — that varied, but invariably busy process of making many rapid judgements and decisions about a host of assorted matters as the working day progresses. Despite giving space to the treatment of this topic in Chapter 1, where it served to introduce discussion of some of the main ideas within the new cognitive science, we agree with that long tradition in administrative studies that rates decision-making as of prime importance to any understanding of organizational life and its many sustaining practices. Hence our desire to develop the topic further, and in particular, to sketch some practical applications of earlier neural network ideas for decision-making.

## Norms and Choices

Possibly the most influential approach to theorizing decision-making in the last fifty years is that body of theory or, perhaps more accurately, that research program referred to as normative decision theory. As its title implies, the broad aim of the approach is to specify what decision *ought* to be made in a given situation of choice. By 'ought' is meant what should be done according to some principle, or model, of rationality. (See Heap *et al.* 1992, for an overview of many features of the program.) Although its methods are obviously heterogeneous, owing to the immense variety of decision problems, something of the flavour of the approach can be seen with the aid of examples. Consider, first, a simple game of chance played with two unbiased dice with faces marked 1 to 6. Suppose that the aim is to guess the numerical value of the combined face-up numbers after a throw of the dice. Possible totals therefore range from 2 to 12. Suppose further that a US$10 bet on a guess will, if successful, return US$80 dollars, regardless of the number guessed. What one ought to do, in this situation, is bet on a guess of the total being 7, because out of the 36 possible ways in which the dice can land face-up, 6 ways will yield a total of 7. Since 6 ways out of 36 means that one can expect the total to be 7 every 6 throws, each investment of US$60 will, over the long term, give a return of US$80. Consistently maintained, this guess will turn the game into a money pump from the 'house' to the player.

Within this small universe, the probability of all throw outcomes is known, and the value of the player making money is assumed. Unfortunately, such abstractly specified games never actually take place in abstractly specified circumstances. Set in the context of a fundraising function for a school, a player who is a pupil's parent may be alarmed at the 'house' losing and draw attention to the incorrectly configured pay-off odds. The assumed player value attached to making money, to be generalized beyond the example, needs to be amended to state, perhaps, that it is of value to make money, except when conditions C obtain, with these conditions to be specified either on a context by context basis, or in terms of a set of generally formulated over-riding values that form some kind of structure. The upshot, however, is that the simplified example now offers no guidance on what ought to be done until these contextual factors are added in. (See Chapter 7, for a similar difficulty with rule-based accounts of morality.) A school fundraising event and a big city casino are sufficiently unalike to provoke a reshuffling of player responses.

Despite differences in contexts revealing differences in priorities among a person's values, the normative tradition's dominant view of rationality, namely maximizing expected value, still operates. It's just that the *structure* of a person's values may be complex. Consider another small universe where this normative principle of reason can be seen to operate, a universe where you are deciding whether to bring an umbrella to work. The elements of the decision problem are summarized in Figure 6.1.

The figure should be read as follows. The probability that it will rain is taken to be 0.5, or 50%, and hence the same as that for no rain. On the other hand, the value of carrying an umbrella, or not, will vary depending on whether it rains or not. There are four possibilities. To bring it, with rain eventuating, has high value (+6) because of the advantages associated with staying dry. If it does not rain there is a small disadvantage (−2) caused by the inconvenience of carrying an unneeded umbrella.

**Decision-making under risk**

|                 | Rain      | No Rain   |
| --------------- | --------- | --------- |
| **Bring Umbrella** | + 6, 0.5 | − 2, 0.5 |
| **Leave Umbrella** | − 8, 0.5 | + 2, 0.5 |

**Figure 6.1** Should I carry an umbrella to work? The number to the left in each cell is the utility for each alternative. The number to the right is the probability of rain or no rain.

The decision not to bring it, and rain occurs, is disastrous (–8), while if it does not rain there is the modest advantage (+2) of not carrying around something unnecessary. As the theory of reason counsels maximizing expected value, we can put all this into a calculation of expected value for each alternative and make the appropriate choice as follows:

*Expected value of bringing the umbrella:*      $0.5 \times 6 + 0.5 \times (-2) = 2$
*Expected value of not bringing the umbrella:*      $0.5 \times (-8) + 0.5 \times 2 = 3$
Therefore, one ought to carry an umbrella.

In the real world, the probability of rain will vary each day, though presumably it will be given by the weather bureau. But what factors contribute to the setting of the numerical values associated with outcomes? Again, the usual contingencies prevail to influence even one's ordinal rankings. In risking a downpour, things like the temperature, how well you feel, will you have the time and convenience to dry out, what clothing is worn, is there space in your bag, on this particular day, for an umbrella to be conveniently carried, and so on, will contribute to a value flux. Notice, also, that valuation depends on the calculation of possible consequences of decisions, and hence contains a nest of embedded probability judgements. So, how bad being soaked is judged to be, will depend on what you think will happen.

If judgements of value are thought to depend on estimates of probabilities, scenarios and likely outcomes, then it is also true that these latter items are influenced by values. Extensionally equivalent expressions, that is, expressions that refer to the same thing, can be valuationally divergent, creating what is known as framing effects. (Goldstein and Weber 1997, pp. 581–582.) Being known as 'the most academically successful school in the state' and being known as 'school number 1131' will lead parents to make strikingly different estimates about their children's potential school success, even though the expressions refer to the same school. Once estimates of described states of affairs are seen to depend, not on referential properties of those descriptions but on intensional, or meaning-related properties, a person's global theory comes into play, as meaning is partly dependent on the conceptual role of an expression (or concept token, to avoid an excessively linguistic formulation in a person's theory.)

The devil is in the detail. It is not in question that powerful normative decision procedures can be specified for certain specific, mostly well-structured, problems. Whole industries — insurance, for example — demonstrate this every day. The difficulty is in scaling such procedures up to the point where they might be useful for a more complex universe. For the above two examples, this complexity does not yet include explicit mention of a social world. Interestingly, normative theory has been extended into the social realm, by way of simplified models of social interactions called games, where…

> …[a] game is defined as a situation in which the actions of one person perceptibly affect the welfare of another and vice versa. These effects can be

classified according to the degree to which there are motives for cooperation and for rivalry. … Whatever the case, the basic method of game theory is to argue that individuals try to predict what others will do in reply to their own actions, and then to optimize on the understanding that others are thinking in the same way. (Heap *et al.* 1992, p.94).

It is the act of optimizing that has been developed to function normatively. The landmark work in the field is von Neumann and Morgenstern's (1944) *Theory of Games and Economic Behavior*, which dealt with cooperative games. The basic work, in extending the theory to non-cooperative games, was done by John Nash (1950, 1953). As non-cooperative games model the most frequent social situations for decision, we illustrate some key ideas with an example.

Consider a school in which government funding cuts have led to a deterioration in teachers' work conditions. As a result, the local teachers' union has imposed a ban on all school activities not directly related to teaching. The Principal, on the other hand, sees the ban as restricting the school's capacity to promote itself in the wider community through the now denied opportunities of school concerts, plays, and other public displays of school performance and achievement. The pay-off matrix in Figure 6.2 describes the essential features of value, or utility, gains and losses for each of the main actors: the Principal and the Union Leader.

In the four cells, the first number of each pair represents the Principal's (*P*) expected value (or utility), the second, the Union Leader's (*UL*), for each of two alternatives. One alternative is for each party to weaken their demands and shift to a compromise set of proposals. The other alternative is to maintain a strong set of demands and resist any attempt at compromise. If both *P* and *UL* maintain their conflicting demands, eventually there will be a damaging strike and the school, which suffers from falling enrolments, will experience an acceleration of this trend and will be forced to close, resulting in a substantial disutility (–8) for both *P* and *UL*. If *UL* makes major concessions and compromises while *P* continues to resist, the result for *UL* will be a substantial loss of authority, respect and power, reflected in a large disutility (–10), while *P* will gain (+2). Similarly, if *P* concedes, and *UL* remains firm, the utilities will be reversed. But if they both compromise, there is only a modest loss for each (–1). In conflict situations where the greatest personal gains are to be found in victory over an intransigent opponent, where equal intransigence is disastrous, and where equal compromise produces only minor personal losses in prestige and authority, this simple design is commonplace. The pay-off matrix in Figure 6.2 is also equivalent to that constructed for a very well known problem in decision theory, namely the Prisoner's Dilemma. The puzzle develops when each participant pursues an individually rational course of action.

Watch what happens in a pseudo-bargaining situation, where each participant is calculating utilities based on making the same assumptions about the other person's calculations. (We say 'pseudo-bargaining' to indicate that even though communication is going on, the important role of bluff compromises knowledge of truth.)

**Union Leader**

|  | Compromise | Resist |
|---|---|---|
| **Compromise** | - 1, - 1 | - 10, + 2 |
| **Resist** | + 2, - 10 | - 8, - 8 |

**Principal**

**Figure 6.2** Pseudo-bargaining in a school: a Prisoner's Dilemma conflict. The number to the left in each cell is the utility for the Principal.

Suppose *UL* resists (utility = +2). Then *P* ought to resist because the utility of resisting (−8) is greater than compromise (−10). Suppose now that *UL* compromises (utility = −1). Then *P* ought to resist because resistance (+2) has greater utility than compromise (−1). Therefore, whatever *UL* does, it is rational for *P* to resist. Now because the pay-off table is symmetrical, the same reasoning can be advanced on behalf of *UL*. That is, whether *P* compromises or resists, it is always better for *UL* to resist because in each case *UL*'s utility is higher (giving values of +2 and –8 for each respective option). The puzzle, of course, is that in both resisting, the result is considerably more disutility for each than would have been the case if they had both compromised (−8 as opposed to −1). The reason it is a puzzle is because, construed as a normative guide, as an account of what ought to be done, the pay-off matrix and simple reasoning rules grind out a result where both participants end up in a situation whose utility is described by the bottom right-hand cell of the matrix (−8, −8) rather than, say, in the much better situation corresponding to the top left-hand cell (−1, −1).

The literature on the role of game theory in this common circumstance is voluminous, (see Skyrms 1996) so we shall make our comments correspondingly brief. Provided rationality is defined very narrowly — in this case the reasoning is 'minimax' because each opponent seeks to minimize the maximum loss the other can impose — there is no obvious alternative strategy for players to adopt to avoid this outcome. Accordingly, the outcome is said to reflect a *Nash-equilibrium* point. But as Ross (1996, p. 1) remarks:

> Nash equilibria are stable, but not necessarily desirable; for example, in what is undoubtedly the best-known and most-discussed instance of a game, the Prisoner's Dilemma, the unique Nash-equilibrium is a state in which both of the two players are as badly off, given their utility functions, as possible.

In view of the normative uses envisioned for game theory, it is ironic, therefore, that Ross (1996, p. 1) should immediately infer: '[t]he point of Game Theory, then, is not prescriptive but descriptive...'

While congenial, this conclusion nevertheless seems to us to be a bit too strong. Our point is that the normative value of a decision-making procedure is something that needs to be argued for, possibly on a case by case basis. Note, however, that when it comes to educational decision-making, troubles over the normative claims of decision circumstances that are formally equivalent to the Prisoner's Dilemma, are particularly extensive. For example, in Figure 6.2, the decision dichotomy compromise/resist can be recast for a wide class of situations as the dichotomy 'contribute to the common good'/'behave selfishly'. This is clearly the case in policy decisions where education is construed as a public good. As is evident from our earlier remarks on non-social cases, we think that in order to deal with the scaling problem — the problem of shifting from a small artificial model universe into the real world — more global notions of rationality are required which, in normative contexts, invoke the epistemic virtues of coherence. We also think that included here is a need to move beyond individual rationality to socially distributed rationality.

## Decision-Making as Constraint Satisfaction

Our favoured way of moving towards an approach to decision-making that can accommodate both real-life complexity and real-time cognitive processing by fallible human agents, is to take a cue from our epistemology, especially its features of coherence and naturalism. In Chapter 1 we identified a number of coherence criteria for adjudicating among competing theories; criteria such as simplicity, comprehensiveness, consistency, empirical adequacy, explanatory unity, and fecundity. (These are elaborated in Chapter 9, where their role in research methodology is discussed.) In making a decision about what theory to adopt, these criteria function as a set of multiple soft constraints. The theory to be preferred is the one that satisfies the most constraints. If we have a reasonable grasp of the methodology of constraint satisfaction then it can be used as a helpful tool for analyzing decision situations and making decisions.

Many everyday, practical matters can be recast in the form of constraint satisfaction. For example, your school may wish to conduct a professional development function for staff. But what exactly? Something on the latest curriculum innovations perhaps, our old friends school discipline and classroom management, the new funding model and its implications, developing a school charter? As well as the constraint of what people want, we need to consider costs, availability of any guest speakers, the topic's place in a larger sequence of professional development activities, what parents might be expecting teachers to learn, school politics, the goals of professional development, and so on. In general, the best decision will be the process outcome that coheres best with all the relevant constraints.

For many examples, the identification of constraints, and the decisions that are a best fit, will be relatively trivial. Sometimes, however, it is not, demanding more disciplined understandings and procedures. In educational administration, these matters have been analyzed with greatest effect by Viviane Robinson, who has developed a powerful model for dealing with the problems of school practice. (See, for example, Robinson 1993, 1994, 1998.) Although formulated in terms of problems, the account readily goes over into decision-making. Drawing on Thomas Nickles's (1981, p. 111) constraint inclusion account of problem-solving where a problem is defined as '...a demand that a certain goal be achieved plus constraints on the manner in which the goal is achieved...', for Robinson (1993, p. 25), who quotes Nickles's definition more fully, '[t]he process of problem-solving *is* the process of setting or discovering solution constraints'. She adds that '[p]ractitioners draw on past experience, background theory and feedback from the environment to establish and revise a constraint structure'. (Robinson 1993, pp. 25–26) However, in Robinson's formulation of constraint identification, the emphasis is on relations of inference among linguistic expressions, with a strong focus on the practice of dialogue. Procedures and explanations are therefore conducted at a level higher than our own inquiry is examining. For us, the other main set of constraints on understanding decision-making (and also, reflexively, the nature of constraints) is our naturalism.

The specification of coherence criteria, used earlier in our account of decision-making, is couched in terms of adjudicating linguistically expressed bodies of knowledge. But because much decision-making 'in the wild', in naturalistic settings, would be done by human brains processing patterns of visual, aural, olfactory and tactile inputs to sensory organs, the notions of coherence and constraint should have some equivalent realization in the operation of neural networks. In what follows, we pursue two ideas, highly related, to be found in the literature of the new cognitive science. The first is a standard view of constraints embedded in the canonical connectionist model of the natural, or physical, mind. (Rumelhart 1995) The second is a view of coherence itself as something to be understood in terms of such a physicalist model of constraint satisfaction. (Thagard and Millgram 1995; Thagard and Verbeurgt 1998; Thagard 1998.)

In his description of the basic design of neural networks, Rumelhart (1995, pp. 146–148) identifies a number of features of what he calls 'the canonical connectionist model', features that are found in the most basic architectural arrangements for most artificial neural networks. Our characterization of these features, made with reference to the network design shown in Figure 6.3, though closely following Rumelhart's, is in terms of the terminology we have been using throughout.

1. The nodes (or artificial neurons) represented by dots in the figure, stand for 'hypotheses' about the world.
2. The level of activation for each node stands for degree of belief, or 'confidence', in the truth of that hypothesis. In modeling activation levels, the standard mathematical representation is the sigmoid (sometimes called logistic) function given, in its simplest form, by the formula: $a_i = 1/(1+e^{-N_i})$, where $a_i$ is the level of activation for the ith node and $N_i = \Sigma w_{ij} a_j$ which is the sum of the weighted input to that node from each connected node in the previous layer.

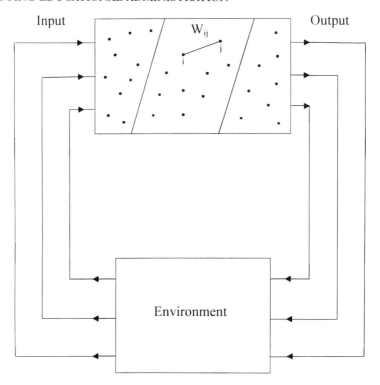

**Figure 6.3** The basic connectionist model. (Adapted from Rumelhart 1995, p. 145.)

3.  *Constraints* are represented by the *weights*, $w_{ij}$, between two nodes. If the weight is large, it means that there is a large constraint between the connected nodes. A positive large weight means that if the hypothesis represented by one node is true then it is likely that the hypothesis represented by the other node is also true. A negative large weight, also a strong constraint, signals the opposite: the truth of one hypothesis being most likely associated with the falsehood of the other. Weights in value closer to zero indicate weak, or no constraints.
4.  The net's inputs function as its empirical evidence for what the world is like, its outputs constitute its 'action' on the world. Inputs could thus be thought of as ways of setting the activation levels of the first layer of well chosen nodes, or hypotheses, by being construed as confirming or disconfirming empirical evidence.
5.  When the process of weight adjustment ceases, for some defined cycle of inputs, the net is said to have *settled into an interpretation* of the input patterns. (Rumelhart 1995, p.147.)

How do these basic features link to our preferred coherentist epistemology, and how can they be developed to characterize decision-making?

We draw on the work of Paul Thagard, especially his notion of decision-making as constraint satisfaction, to take our position further. Consider, first, the epistemological issue of maximizing the coherence of a large body of statements, S, or linguistically expressed propositions. As the demands of consistency, comprehensiveness, deducibility, explanation, and the like begin to be applied, we shall discover that the elements of $S$ can be partitioned into two groups, $T$ and R, such that $S = T+R$. The first group, $T$, contains statements that *cohere* with each other. The second group, R, contains all those statements that *incohere* with the elements of $T$. (Thagard and Verbeurgt 1998, p. 3.) For Thagard, who wishes to define more precisely the conditions under which $S$ can be partitioned into these two groups, the sense of 'constraint' defined in Rumelhart's canonical model, lends itself to a natural elaboration. Constraints represented by positive weights between the elements of S are said to indicate that these elements *cohere*, whereas negative weights indicate that the elements *incohere*. (Thagard 1998, p. 407; Thagard and Verbeurgt 1998, p. 3.) Weights that indicate coherence among elements could be expected to represent the usual candidates — for example, consistency, deducibility, or explanatory unity — whereas high negative weights, indicating incoherence, would code for contradiction, or incompatibility.

Now from an epistemological point of view, a good theory will have high positive constraints with some claims and high negative constraints with others. For example, in administrative studies, the hypothesis that human nature is basically good, will cohere with that cluster of claims associated with the tenets of Theory $Y$ asserting the essentially cooperative nature of organizational life and its corresponding examples of evidence of altruistic behavior, and incohere with the claims of Theory $X$ and its less generous view of human nature. Or, to use Thagard's (1998, p. 409) example:

> …Darwin's theory of natural selection consists of propositions that have positive constraints with biological evidence and negative constraints with creationist hypotheses; maximizing coherence requires accepting Darwin's theory and rejecting creationism.

On this approach, maximizing the coherence of $S$, amounts to maximizing constraint satisfaction, which includes both positive and negative constraints. However, the *best theory* that emerges from this learning process is, in terms of the model, given by the nodes that have the highest activation levels. On our understanding of coherence justification, this would be the hypotheses partitioned into $T$, the set that contains all the positively constrained higher activation levels. Such a consequence follows from the canonical practice of interpreting activation level as degree of confidence in the truth of an hypothesis, and the relative closure of this set under the partitioning relation of constraint satisfaction.

As we accept a coherence theory of *evidence* and a correspondence theory of *truth*,

(Evers and Lakomski 1991, pp. 42–44) one may be puzzled over how we sanction a move from degree of confidence to truth. Essentially, our point is that the hypothesis of an external world in causal interaction with our practices of experiment, inquiry and attempts at successful navigation, coheres positively with our best coherently justified theory. (See also Thagard and Verbeurgt 1998, p. 16.) Of course, on Thagard's model, hypotheses are localized at particular nodes rather than being distributed across nodes as we suppose is more biologically realistic. Nevertheless, realism in model design comes in grades, and simplifications can be powerfully instructive and revealing, even to the point of showing how to support the relationship between evidence and truth.

With this conceptual apparatus in hand, we now have a way of dealing more effectively with the global features of rationality that were problematical on traditional normative approaches to decision-making, especially those in the game-theoretic genre. We use an example based on the work of Thagard and Millgram (1995).

You have been Vice-Principal of a good school for several years, and have recently been offered the chance to move to another state and be Principal of an excellent school, paying more money. There are several advantages. The job offers more prestige, excitement, greater professional challenges and, of course, the extra salary. On the other hand, the move would create considerable discomfort for your family, and also be disruptive to your own personal life. In going beyond these specific goals to reviewing your broader goals, you note the value you attach to keeping your family happy, keeping yourself happy, and making a contribution to education through your career. Against this mixed set of considerations, your decision concerns which action to engage in — either to accept or reject the job offer.

Presenting the problem this way, highlights two significant aspects of most decision situations. First, although the list of relevant considerations is not large, it gives some indication that, in practice, there is always a diverse range of issues to be taken into account, and that in accounting for them successfully, decision methods should be able to scale upwards to deal with an even larger range of issues, without drastic refinement. And second, the considerations are mixed in the sense that some of them cohere, or clump together, and others are opposed, or even contradictory. One way in which goals and actions (called factors) can cohere is through a process known as facilitation, meaning that some factors help bring about others. (Thagard and Millgram 1995, pp. 441–447.) Assumptions about facilitation provide information about positive constraints. Factors can also incohere, or be incompatible. Thus, both actions — accepting or rejecting the offer — are mutually exclusive and therefore negatively constraining. When all of this information (plus data for some additional matters to be explained below) is coded as input to a neural network simulation program that attempts to maximize the coherence of all the factors (or hypotheses represented by the network's nodes) by the process of adjustment towards maximum constraint satisfaction (in this case, Thagard's DECO program) the result can be represented by Figure 6.4.

One lot of additional information required by the model is a way of indicating the priority of the broader goals, $G1$, $G2$, $G3$. The label 'ACTIVATION' signifies that

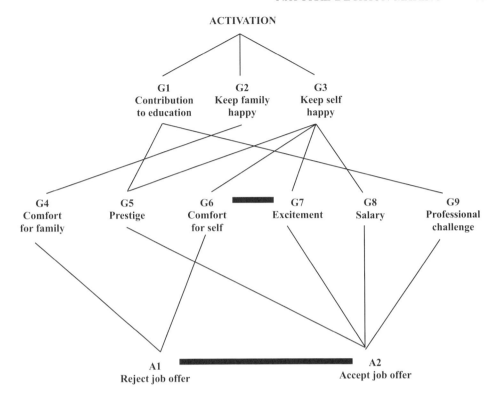

**Figure 6.4** Decision network. The thin lines mean the constraints between factors are positive and facilitating. The two thick lines signify that the constraints are negative and incohering. (Adapted from Thagard and Millgram 1995, p. 448.)

a steady source of activation is applied to these nodes via positively constraining weights. In a range between 0 and 1, the activation level can be set at 1. Initially, all other goals can be set at zero, although initial weights must be set for all the facilitating and incompatible links. In the similar decision example that Thagard models, all facilitating weights are set at 1, and of the inhibiting weights, the link between *A1* and *A2* is given an inhibition strength of 1, while the weight between *G6* and *G7* has an inhibition level of 0.5.

With this information, the artificial neural network is able to begin cycling it around, making a series of mutual adjustments that will result in a particular set of activation levels and a maximized set of weights when the network has eventually stabilized. Notice that in this kind of model, the flow of activation proceeds from the top, where the broad goals are located, down through to the two actions. Normally, when people are trying to make a decision, they appear to proceed upwards, from each action to a calculation of consequences. And what is the outcome of the network's deliberations? Well, the simulator decided to take the job. *A2* ended up with a high level of activation, while *A1* ended up being deactivated. However, if *G4* and *G6* had been broad priority goals, instead of *G1*, *G2*, and *G3*, the net would have decided *A1*, and rejected the job offer.

Such is one important way of using developments in the new cognitive science to develop a better, more realistic understanding of decision-making. In this case, the issue being addressed is the holistic nature of deliberation, something which we saw caused difficulty for traditional decision-making normative models with their means-ends focus on rationality. Decision-making that proceeds by attempting to satisfy multiple soft constraints is, in our view, a better approximation to what is required of human decision-makers in everyday naturalistic settings. Moreover, as can be seen from a previous discussion of Hutchins's (1991, 1996) work (in our Chapter 5) the canonical connectionist model, with a localized representation of hypotheses as nodes in a network, lends itself to modeling group processes, where individuals are represented by small networks and the group consists of these nets being in turn linked by variously constraining weights. Modeling the effectiveness of group learning by exploring the consequences of adopting different values for weights, throws light on the design of organizations that learn, and also that make decisions.

## Dealing with the Non-Linguistic

The localized representation of knowledge in nets has a number of advantages. It does not stand in the way of instantiating holistic, constraint satisfaction based reasoning, and it permits ready access to the coding of familiar decision issues. However, the second advantage is mostly due to the availability of suitable linguistic formulations for the hypotheses located at activation points or nodes. But for much decision-making in the wild, it is more likely that the distributed representation of visual scenes, or sounds, in the form of input *patterns* is the norm. As yet, there is little of practical value in the literature for modeling these kinds of inputs in a decision-theoretic useful way for practitioners in naturalistic settings. The example of medical diagnosis discussed in Chapter 1, does use distributed representations of identified symptoms, but this is after a coding regime had been established for converting high level linguistic/symbolic expressions into a suitable vector format for inputting to the net's first layer.

Given our emphasis on the role of non-linguistic patterns as a form of decision information presentation, and the importance we attach to attending to the *processes* of information usage, notably the study of systems that *learn,* we would like to conclude this chapter with a brief discussion of some of the work coming from researchers in the 'naturalistic decision-making' tradition. (See Zsambok and Klein 1997, for an overview of the approach and a good sample of the issues it deals with.) In focussing on the problem of how to *teach* decision-making, particularly on how to train novices to be more like experts in making decisions in naturalistic settings, we find this work coheres with much of our own theoretical orientation derived from epistemology and cognitive science. Indeed, the main obstacle we see to offering a fully naturalized account of practical decision-making is the gap between the ubiquity of pattern recognition as decision inputs in everyday contexts and localized representation of decision knowledge in ANNs.

We begin with a discussion of a decision training problem canvassed by Klein (1997). (A fuller account of our view of training is contained in Chapter 8.) The problem concerned the identification, by nurses, of systemic infection (sepsis) in babies in a Neonatal Intensive Care Unit. In this decision context, judgments need to be made rapidly, under pressure, with ambiguous and subtle cues, in a high risk situation. A study of very experienced neonatal nurses found that they relied heavily on perceptual-recognitional skills. (Klein 1997, p. 345.) The study also found that they had difficulty in articulating what they were recognizing. Since the perceptual factors that were relevant were not, at the time, clearly identified in the nurse training literature, this is probably the kind of skill that developed under pressure of feedback from perceptual cues, complex patterns of visual, oral, tactile, and olfactory sensations that, over time, were matched with patient outcomes. To shorten the process of training for novice nurses, the experts were carefully interviewed in an effort to determine precisely the nature of these patterns and cues, a kind of coding process that translates non-linguistic perceptions into linguistic expressions of what, of significance, was seen. After noting all the convergences of expert identified cues, the linguistically formulated schedule of cues then formed the basis of a program of training in perceptual pattern recognition for novice nurses. So even though the symbols may have failed adequately to compress the experience of the experts, they served the purpose of directing attention to salient features of the passing show of experience. In short, they helped prompt the perceptual recognition process in a diagnostically more fruitful direction. When trained on what to look for there was a dramatic improvement in the performance of the novice cohort. According to Klein (1997, pp. 346–347), '[t]he accuracy of performance on recognition tests increased from near chance to 87%. The students listed more cues and patterns, and the proportion of these that were accurate also increased'.

Unfortunately, although linguistic formulations of salient patterns and cues served mainly as a locus for promoting the learning of context specific observational skills, the process still relied on the availability of some form of partially valid linguistic coding of experts' accumulated practical knowledge. While this reliance is often reasonable, and can be met with the required formulations, the interesting question, for us, is whether the learning of decision-making can still be accomplished with less dependence on the symbolic representation of decision inputs, outputs, and skills. In addressing a related set of issues from within the naturalistic decision-making tradition, Klein (1997, p. 348) identifies a number of strategies that have been useful in achieving expertise in decision-making. We note several, including some that receive further elaboration in Chapter 8.

First, it is important to engage in deliberate practice, actually to *do* those things that experts do in the course of acquiring their expertise. Going through the motions, as it were, may not seem helpful, but it will shape recurrent pathways of neural learning that will later cohere with further skilled practices. Have in mind goals and evaluation criteria whilst engaging in these practices. For example, decisions on how to improve one's golf swing can be shaped by driving practices animated by the goal of getting the ball into the cup, and data on where it actually lands. Classroom management decisions are shaped by the goal of creating an effective learning environment, and data on classroom behaviour.

Second, 'sample alternative task strategies' (Klein 1997, p. 348). This simply counsels that different approaches be tried out to determine what might work best in a given situation. Again a case of learning by doing.

Third, and something well known among chess Grandmasters, build up an extensive data bank of strategies, techniques and approaches, for different applied contexts.

Fourth, cultivate and promote accurate feedback on the consequences of decision actions. Non-linguistic feedback, for example, can powerfully change the direction of subsequent learning through its influence on future actions performed.

Fifth, engage in reflection. This need not involve the coding of practices into linguistic diary entries, as some using the Reflective Practitioner model have supposed. It is simply to acknowledge that in the time constrained rush of the working day, the best opportunity for further learning might occur afterwards, when scenes are brought to mind, images of sequences of events examined, and alternatives explored in the imagination. Building mental models is suggested as a way of augmenting reflection.

And lastly, obtaining coaching can be valuable for some decision skills. The definition of coaching can be broad, ranging from someone to provide feedback and advice on fine motor performance, to mentorship programs with an emphasis on direct instruction.

In decision learning circumstances, these strategies cohere with a view of practical knowledge representation and dynamics that emphasizes the patterned nature of decision inputs, the contextual dependence of patterns, actions and consequences, the importance of learning through feedback, and the desire to achieve a global coherence of thought and action.

# References

Evers C.W. and Lakomski G (1991). *Knowing Educational Administration*. (Oxford: Pergamon).

Goldstein W.M. and Weber E.U. (1997). Content and discontent: Indications and implications of domain specificity in preferential decision making, in W.M Goldstein and R.M. Hogarth (eds.) *Research on Judgment and Decision Making*. (Cambridge: Cambridge University Press), pp. 566–617.

Heap S.H. *et al.* (1992). *The Theory of Choice: A Critical Guide*. (Oxford: Blackwell).

Hutchins E. (1991). The social organization of distributed cognition, in L.B. Resnick, J.M. Levine, and S.D. Teasley (eds.) *Perspectives on Socially Shared Cognition*. (Washington, DC: American Psychological Association).

Hutchins E. (1996). *Cognition in the Wild*. (Cambridge, MA: MIT Press).

Klein G.A, *et al.* (eds.) (1993). *Decision Making in Action: Models and Methods*. (Norwood, NJ: Ablex Publishing).

Nash J.F. (1950). The bargaining problem, *Econometrica*, **18**, pp. 155–162.

Nash J.F. (1953). Two-person cooperative games, *Econometrica*, **21**, pp. 128–140.

Robinson V.M.J. (1993). *Problem-Based Methodology: Research for the Improvement of Practice*. (Oxford: Pergamon).

Robinson V.M.J. (1994). The practical promise of critical research in educational administration, *Educational Administration Quarterly*, **30**(1), pp. 56–76.

Robinson V.M.J. (1998). Methodology and the research-practice gap, *Educational Researcher*, **27**(1), pp. 17–26.

Ross D. (1996). Game theory, in *Stanford Encyclopedia of Philosophy*. (Website: http://plato.stanford.edu/entries/game-theory/).

Rumelhart D.E. (1995). Affect and neuro-modulation: A connectionist approach, in Morowitz H. and Singer J.L. (eds.) *The Mind, The Brain, and Complex Adaptive Systems*. (New York: Addison-Wesley). pp. 145–153.

Skyrms B. (1996). *The Evolution of the Social Contract*. (Cambridge: Cambridge University Press).

Thagard P. (1998). Ethical coherence, *Philosophical Psychology*, **11**(4), pp. 405–422.

Thagard P. and Millgram E. (1995). Inference to the best plan: A coherence theory of decision, in Ram A. and Leake D.B. (eds.) *Goal-Driven Learning*. (Cambridge, Mass.: M.I.T. Press), pp. 439–454.

Thagard P. and Verbeurgt K. (1998). Coherence as constraint satisfaction, *Cognitive Science*, **22**(1), pp. 1–24.

von Neumann J. and Morgenstern O. (1944). *Theory of Games and Economic Behavior*. (Princeton: Princeton University Press)

Zsambok C.E. and Klein G.A. ( eds.) (1997). *Naturalistic Decision Making*. (Mahwah, NJ: Lawrence Erlbaum).

# 7

# Ethical Practice

In pursuing our theory of practice into the realm of ethics, we encounter two major difficulties. The first concerns the fact that a naturalistic account of practice coheres best with a naturalistic view of ethics. However, a number of influential writers in educational administration claim there are serious objections to naturalism in ethics. We therefore spend the first part of this chapter outlining more general epistemological and semantical reasons for our stance on ethical knowledge. The second difficulty arises out of the importance we attach to particularity and the need to allow for contextual factors. On our theory, the evaluation of practices, the question of determining what is to be done and what is appropriate to do, is sensitive to a theory of the agent. That is, being a teacher, or being an administrator, or being a student can make a difference to the determination of how a person should act. Because we have articulated features of our theory of leadership elsewhere in the book (Chapters 4 and 5) we focus on *ethical* leadership practices, thus constraining the theory of ethical practice with the requirement that it cohere with our approach to educational leadership. Some comments about the generalizability of this framework will be made at the end.

## Naturalism and Holism in Ethics

Several features of our approach to knowledge and its justification have a direct bearing on how we view ethics. The first feature, and one that will serve as a point of entry to others and their consequences for ethics, is what is known as semantic holism. According to the strand of logical empiricism most influential in the development of educational administration, cognitively significant claims are either analytic or synthetic. Analytic claims, such as tautologies, or logical truths, are true by virtue of the meanings of their terms, and hence their justification can proceed without explicit appeal to empirical evidence. Synthetic claims, on the other hand are empirical, but also exhibit this dual nature of evidence in meaning and justification. The specific range of sensory experience that serves as evidence for justifying a particular claim serves as part of an account of the basis for its meaning. Articulated systematically by A.J. Ayer in *Language, Truth and Logic* (1946) as the

Principle of Verification, the meaningfulness, or cognitive content of synthetic propositions, was identified with the range of observations that would either provide evidence for truth or evidence for falsehood. (See Ayer 1946, pp. 35–41). Applying this standard of meaningfulness to ethics, Ayer (1946, p. 108) explores

> ... why it is impossible to find a criterion for determining the validity of ethical judgments. It is ... because they have no objective validity whatsoever. If a sentence makes no statement at all, there is obviously no sense in asking whether what it says is true or false.

Observation is relevant to what is or is not the case, to the empirical facts. For Ayer, and many other traditional empiricists, ethics ends up being merely about our *attitudes* to those facts, expressions of feeling that do not say anything or make a statement.

One reason this view became so influential in the development of educational administration during the 1950s and 1960s was because it was given expression by Herbert Simon in his account of values in *Administrative Behavior* (1946) and played a major role in his conception of a science of administration and his analysis of the structure of decision-making. As we have seen, however, the relationship between observational evidence and the claims it is meant to justify is much more complex, with the principal difference concerning the holistic nature of justification. For as we now know, thanks to the early work of W.V. Quine (1951, 1960), there are at least three important reasons why observational evidence distributes its empirical support globally, across an entire theory, rather than statement by individual synthetic statement as logical empiricism assumed. (See Chapter 1 and also the extended treatment in Chapter 9.) First, observations are theory-laden, being made from the vantage-point of some perspective or another. More technically, the observational sentences that figure in justificatory inferences always contain some theoretical terms. This thesis is of a piece with the claim that all experience is interpreted experience, filtered through the mind of the perceiver. Second, the claims our theories make about the world, because they are couched in general terms, always outrun any available observational evidence. Theory is always underdetermined by experience. This means that the same observations can verify different theories. Finally, because the empirical testing of claims is complex, we have some methodological choice in relocating the blame for an alleged empirical failure away from the hypothesis being tested and onto some other assumption in the complex of background conditions required to conduct and interpret the test.

Because the resulting shift towards holism in justification spills over into a corresponding holism about meaning, traditional empiricism's way of specifying cognitively significant claims, indeed of drawing the cognitive/non-cognitive distinction, will no longer work. With the aid of some further philosophical machinery, we conclude that ethical statements can be as true or as false as any paradigm empirical statement, and as meaningful in exactly the same sense.

Some writers have seen holistic considerations as ushering in an era of subjectivism, because they seem to detach altogether questions about the merits of theories from the once assumed secure foundation of observational evidence (Greenfield and Ribbins 1993). In reply, our argument has been that we need to use criteria of theory choice that are more suitable in these circumstances: criteria that make use of epistemically normative holistic notions such as coherence. When it comes to ethics, we claim that using these criteria permits the justification of ethical statements to be conducted in the same way as for other statements. So, just as empirical assertions are justified by considering the coherence their inclusion confers on a wider system of assertions, we think that considerations of system are equally relevant in assessing the warrant of moral values. (See Churchland P.M. 1989, pp. 297–303; 1996, pp. 91–108.)

To understand the effect of theoreticity in making observations, consider a number of cases. An expert's view of an X-ray photograph is mediated by a substantial amount of medical theory acquired through experience and training. Fractures, or changes in bone density for example, though hidden from the novice can easily be seen with suitable training. Discerning detail, or even making sense of, observations mediated by scientific apparatus, is likewise a theoretical endeavor, as is identifying features embedded in music, art, or literature. Theory, when construed broadly (as epistemology requires) will permit the construction of culture-laden examples of observation, and the role of interpretation and assumed background understandings behind natural and social events. Thus, describing a sequence of behaviors as 'bidding at an auction', or as 'giving offense to the host', can reveal more cognitively significant content and lead to greater predictive success than the use of rival descriptions.

Now the value-ladenness of observation can occur in a similar way. An account of patterns of behavior as 'murder', or 'unfair', or 'generous', can reflect deep familiarity with a network of interrelated concepts that are essential for a person successfully to navigate their way through the social world, anticipating correctly the responses of others in the course of pursuing their own theoretically motivated goals. Part and parcel of this conceptual network can be systematic positions on social justice, respect for persons, the good life, and many context dependent particular moral judgments. In every case, there is no question that the moral theory being used to drive perceptions of the world around, is also enmeshed in a wider fabric of theory that articulates with practice, with culture, with 'commonsense', and with bodies of knowledge grouped into familiar specialisms such as biology, or mathematics, or politics. We can add that the apparent distance of ethics from the observational data of empiricist confirmation and disconfirmation is a result of the theoreticity of ethics, rather than being due to any purported epistemological distinctness.

One significant limitation on the extent to which ethical statements can be enmeshed into a more global theory whose merits can be determined by the application of coherentist criteria, is the belief that despite the surface appearance of an admixture of fact and value among ethical terms, these two elements are logically distinct and can never be combined. On this view murder, for example, consists of some empirical state of affairs — the factual component — and a separate evaluative component to do with imputing wrongness. Empirical evidence is relevant to establishing the facts concerning a killing, but beyond that moral approbation is left over as a feeling, or an expression of sentiment towards the facts.

In the course of developing accounts of moral justification, a number of writers within the cognitivist tradition of ethics have sought to close this gap. John Rawls in his celebrated *A Theory of Justice* (1971) seeks to enmesh principles of justice within a theory of rationality, of what is presupposed for an agent to act rationally. The basic strategy is to defend these principles by showing they would be chosen by rational persons acting under conditions of impartiality. Habermas (1972) is another writer who employs this Kantian strategy. He seeks to enmesh moral principles in an account of what is presupposed by any act of human communication. Background to each speech act is thought to lie in an Ideal Speech Situation, which includes such imperatives as truth telling, or the right to fair participation in the speech community.

Although the sense of 'ought' that is sought is broader than mere affect in being located in a view of what 'oughts' are embedded in human reason, or communication, we are skeptical of this Kantian enterprise. In general, we think that attempts to establish presuppositions of all thought, or all thought on a particular topic, are in practice attempts to establish the assumptions behind some particular *theory*, perhaps a theory of reason, or a theory of communication. That is, theory-ladenness invests the Kantian enterprise. In Habermas's case, we find the branch of science that deals with communication in an information theoretic way more plausible, with its requirement for statistical regularity rather than truth. And in Rawls's case, maximin reasoning — always hedging against the worst possible outcome — ignores the complexities of rational, context sensitive, risk taking. Normative accounts of human reason cannot be so readily detached from empirical studies of human cognition.

Another influential approach to ethics which attempts to enmesh the ethical and the factual, is hedonistic utilitarianism, which seeks to *define* good naturalistically, in terms of human happiness. However, there has been a long debate in ethical theory about whether any attempt to produce a definition which equates a moral term with a set of non-moral terms commits a fallacy — the so-called naturalistic fallacy. (See Smith 1995, for a guide to the recent literature and debates.) Our position in this debate is on the side of naturalistic definitions (Evers and Lakomski 1991, pp. 166–177). Following Quine (1951) we think that semantic holism can be extended to blur not just the distinction between factual and theoretical synthetic statements, but also the distinction between analytic and synthetic statements. (A key part of Quine's argument appeals to the complexity of test situations where he claims that any statement in a theory can be rendered immune from empirical revision, and hence analytic, by methodologically choosing to revise other statements in the theory.) The reason rejecting this latter distinction is important for ethical theory is because the usual attempts to formulate the naturalistic fallacy argument require that a definition be analytic. For example, if it is assumed that a definition is analytic, then we get G.E. Moore's famous 'open question argument' which runs as follows. If an analytic statement cannot intelligibly be questioned, and for every purported naturalistic definition of 'good' we can intelligibly ask 'But is 'good' really (say) 'the greatest happiness for the greatest number'?', then there cannot be a naturalistic definition of 'good'. (Evers and Lakomski 1991, pp. 169–172 contains a more detailed discussion.)

Once we accept enough of semantic holism to blur the analytic/synthetic distinction (and we emphasize that we do not see conceptual role as the sole arbiter of meaning), we are able to adopt a less restrictive account of definition. For example, in science we may come to accept the definition of force as mass times acceleration ($F=ma$), because of its theoretical centrality in the most coherent global account we can give of certain phenomena. The definition will fail the open question argument, but so what? Its acceptability resides in the excellence of the theory in which it is embedded. Our point is that the same reasoning applies in the case of ethics. A naturalistic definition of 'good', or any other ethical term, will owe its plausibility to the extent to which it figures centrally in a theory of human conduct and behavior that meshes with the most coherent account we can give of experience. If versions of hedonistic utilitarianism are to be found wanting, it won't be because of failure of analyticity. It will be because they fail to organize considerations of human motivation, cognition, judgment and intelligent action in an epistemically satisfactory way.

While any number of different approaches within naturalism may emerge as enjoying more of the advantages of system than rivals, our own approach takes note of the provisional, fallible, and experimental nature of all knowledge, as emphasized by Dewey and Popper. To solve our problems, relate satisfactorily with others in the pursuit of our individual and collective plans, and to navigate our way around the natural and social world with the possibility of success better than chance, there is a premium on good theory. Whatever the final shape of a systematic naturalistic account of ethics, one central organizing feature we assume, following Dewey, will be an articulation of the virtue of promoting the long-term growth of knowledge. We see this as a touchstone virtue in the sense that it is shared by a very wide range of perspectives, and hence is more likely to turn up as a vital part of almost any attempt to build a most coherent global theory. So, instead of appealing to *a priori* accounts of rationality, or communication, in order to understand the conditions under which the growth of knowledge can occur, we appeal to empirical theories of *learning*. The result, we believe, will include some of the ethical aspects of human modes of association relevant to individual, organizational, and social learning.

Maintaining both that knowledge is valuable and that our capacity to know is limited and fallible places a premium on the development of knowledge acquisition processes. But what kind of knowledge acquisition processes do we need to attend to in developing a framework for understanding ethical practice? If our general thesis about the representation of practical knowledge is correct, then a naturalistic construal of ethical knowledge in terms of how the brain stores, processes and changes its information will have important consequences for ethics. (See Churchland 1998) We begin by briefly highlighting three such consequences.

First, for a very large proportion of our knowledge, and most of our day to day practical knowledge, we have no symbolic, language-like representations at all. This knowledge, or know-how, although regular and patterned, cannot plausibly be represented by linguistically formulated rules. Examples include, judging whether the milk has gone off, or when to change gear as the car goes uphill, or whether some purchase would make an appropriate present for someone, or whether this child might benefit from a special curriculum, or a thousand other humdrum to major matters that we deal with every day. And to this list we would add ethical judgment (Churchland 1995, pp. 123–150).

Attempts to codify moral knowledge into a linguistic representation present two key problems which we canvass more fully in the next section. First, the desire for generality leads to a failure to guide in any particular case. At the extreme, to advise 'Do good and avoid evil' is not to advise at all unless some specification of what counts as good and evil is given. But second, giving such a specification makes for particularity, leading to a failure to advise beyond time, place, circumstance, participants, and issue. The usual trade-off is to describe instances at some intermediate level of description and advise to act in such-and-such a way *in relevantly similar situations*. Since the same problem breaks out for the task of specifying relevance and similarity, we suggest that in practice the problem is dealt with by the brain learning to extract *prototypes* of moral situations from experience. Prototypes, which capture the statistically central features of experience, serve as non-brittle templates for the identification of patterns of similarity and relevance in the flux of moral life.

In his political theory, Karl Popper (1945) advocated open, free and democratic societies, as these possessed the requisite features for the organized testing of social theories. Fallible knowledge can only be improved if there exists the machinery for putting its claims to tests. While we think Popper's account of theory revision and error elimination needs to be augmented with an account of how coherence constraints function to select the most epistemically progressive possibilities for revision, we concur with his general point, that the improvement of theory occurs best in the context of a particular set of social arrangements regulated normatively by an ethics whose key features include tolerance of diverse viewpoints, freedom of speech and association, and respect for persons.

From neural network models of individual cognition that we have been advocating, we believe that significant natural learning occurs, including the building up of prototypes, through the processing of feedback according to the coherent application of multiple soft constraints. In light of experience we appear constantly to adjust our inner representation of knowledge in an effort to realize the most reliable path through the possibilities and options presented. Where experience can be summarized into useful linguistic formulations, these are able to go proxy for our own efforts, thus providing an efficient way of learning from others. However, again there is the same premium on an ethical infrastructure for learning: tolerance of viewpoint, openness to criticism and alternatives from one's own, willingness to foster freedom of expression and forms of association that permit the free exchange of ideas.

In attending to the ethics required to promote learning and the growth of knowledge, itself a body of knowledge built up and represented in the brain, we have tended to focus on limits to human cognition. The new cognitive science does, however, suggest some natural strategies for ameliorating these limits. The strategy that is of most importance for the ethics of educational administration is what is known as the externalization of cognition. Limitations to memory can be partly overcome by keeping records — writing things down — and then when this becomes extensive, using a retrieval system (See Chapter 3). More importantly, very difficult, or perhaps very complex learning tasks can be dealt with by distributing the cognitive load socially, as a mode of organization, a point we will revisit later. The possibility of more effective problem solving by resort to organization dovetails nicely with the point that individuals can comprehend patterns in modes of association, locating themselves coherently within such patterns. Having a sense of structure, a sense of neighborhood relations and a grasp of a modest portion of the task seems sufficient to permit the off-loading of much individual cognition onto a mostly external division of cognitive labor.

## Ethical Leadership and Moral Codes

Granted our perspective on semantic holism and naturalism, how do we theorize ethical leadership in educational contexts? On the one hand, some well known approaches to leadership emphasize the importance of moral guidance as part of what is required in being an inspirational leader, one able to transform followers and initiate significant organizational change. (Leithwood, Tomlinson and Genge 1996, p.786.) On the other hand, the ubiquity of uncertainty and the sheer complexity of modern organizational life conspire to compromise the value of knowledge behind any proposals for moral guidance. What is needed is a scheme for integrating the demands of leadership with the constraints that make moral knowledge so difficult to achieve. In tackling this issue the main strategy will be first to outline an approach to representing moral knowledge and then, supposing that a theory of leadership is required to cohere with it, use the coherence constraint to develop the main features of that theory of leadership as it pertains to ethical matters.

Requiring coherence between these two bodies of theory is not unreasonable, especially where large scale theorizing is being attempted. For example, Hodgkinson's (1991) model of leadership in terms of position within a stratified hierarchical organizational structure is based directly on his model of ethics as a stratified hierarchy of differently justified claims. Adequate leadership at the top of the organizational hierarchy requires cognitive access to a special class of values (those that are 'transrationally' justified) at the top of the values hierarchy. (Hodgkinson 1991, pp. 143–165.) And Critical Theory accounts of ethics based on the alleged moral presuppositions of maintaining what is known as an 'ideal speech situation' emphasize the kind of moral principles used to defend democratic and participatory styles of leadership, principles to do with tolerance, equity, fairness and justice. (Foster 1985.) Even the implicit theorizing embedded in cultural practices appears to press for coherent resolutions in these matters. For example, Wong (1996) draws attention to some significant differences between Eastern and Western moral cultures noting, in Confucian thought, the ethical importance of learning in both character development and in the goal of serving the people. But he also observes that these values are consonant with leadership practices that emphasize consensus, group processes, and communitarianism.

One way of demonstrating moral leadership is through the development of a code of practice, to prescribe appropriate conduct by articulating a set of written guidelines or rules. The Ten Commandments are an example of one such set of rules. Many organizations and professional associations with less lofty purposes develop their own distinctive codes. The 'Statement of Ethics' approved by the National Association of Secondary School Principals, in the US, prescribes that the educational administrator, for example: 'Makes the well-being of students the fundamental value in all decision-making and actions', and 'Fulfils professional responsibilities with honesty and integrity'. Actually, these two principles highlight one of the difficulties to be found with ethical codes. Because a code is meant to be applicable in general and across differing circumstances, its statements will be fairly abstract. Perhaps the most general and abstract prescription is the injunction to 'do good and avoid evil'. The problem is that what counts as doing good and avoiding evil is left entirely open. As a result, the statement provides no guidance. Similarly, there will be little controversy over making the well being of students the fundamental value in school administration, but much debate over what counts as student well being, or whether certain particular proposals will effectively promote it.

To increase the use value of codes as moral guides providing a source of moral leadership in the light of the generality problem, it might be thought that some specifics should be included, or that moral principles be made more explicit. An example of explicitness would be 'Always tell the truth'. Unfortunately, without any qualifiers, this statement looks to be mistaken, as it assumes that all people in all circumstances have a right to be told the truth. But the misuse of knowledge can sometimes be a reasonable ground for withholding truth, or even for lying. A person robbing a bank is not automatically entitled to be truthfully informed of the combination to the safe. Explicitness also renders more likely a clash with other moral principles. Concern over hurting the feelings of another may prompt lying about relatively trivial matters: e.g. responding to a question about the appropriateness of a choice of footware with the words 'That's a nice pair of shoes you're wearing'. Two or more explicit, independent, non-trivial moral rules can always be shown to conflict in some situation. Under these circumstances, most of the effort required for moral leadership comes from outside the code of conduct.

The same point can be made when a code's principles are hedged in with written qualifiers, as in 'Always tell the truth except when conditions $C1$, $C2$, $C3$, …, $Cn$ obtain'. Either the qualifiers ($C1$, $C2$, etc.) are quite general, in which case a version of the generality problem will break out again, or they are specific. But the trouble with specifics is that there is no obvious end to them. The hedged statement is what is called 'infinitely defeasible', admitting an open class of legitimate exceptions.

These difficulties in using codified maxims for moral guidance can be characterized more broadly as follows. Inasmuch as the maxims are expressed in general terms, they will derive their force as moral guides through the process of interpretation, a process that lies outside the code. And inasmuch as the maxims are written to try to capture specifics, the particularities of contexts will always outrun the particularities able to be captured in linguistic formulations, thus requiring an external source of guidance to waive some maxims and augment others with missing detail. (For some of these points made in relation to the evaluation of codes of ethics in educational research, see Small 1998.)

## Symbolic Moral Reasoning

Because models of leadership that articulate with the provision of ethical leadership through the development of codes of conduct are relatively open, pending some specification of the code-maker's moral authority, we look naturally to moral theory as a potential antecedent source of guidance. The two most influential models of moral reasoning to be found in the administrative literature are best seen as articulating with the broad decision-making tradition of leadership. This tradition places a heavy premium on representing knowledge as symbolic, linguistic structures.

In applying symbolic representationalist accounts of knowledge to ethics consider, again, classical (or hedonistic) utilitarianism, which may be formulated roughly as the moral principle that one ought to do that which maximizes the total amount of human happiness (or minimizes the total amount of human misery). Typically, rational moral evaluation under this rule is assumed to require a close specification of the circumstances of each action sufficient to yield an empirical estimate of the quantity of happiness that would result. Although the principle can be simply stated, hedonistic utilitarianism places formidable demands on the cognitive powers of anyone using the theory as a source of moral guidance. First, because it requires an estimate of outcomes of unrealized alternatives — it needs to take into account hypothetical courses of action — moral agents would need to possess quite detailed theories of the causal operation of complex social systems such as schools or education bureaucracies. But these theories are simply not available, except in very abstract functionalist versions ill suited for fine-grained causal prediction and analysis. Second, there is a puzzle over whether the only morally relevant outcome is quantity of happiness, or whether quality is also important. John Stuart Mill thought that quality was important. But if so, how are the two to be traded off in a decision-making context? Indeed, with the resources of language, how are we to describe, in measurement-theoretically useful ways, the relevant levels or amounts of both quality and quantity of happiness? Finally, there is the problem of how happiness could be measured at all: how could one ever know what lies behind the often inscrutable behavior of others?

Preference utilitarianism, the form that has been most influential in administrative science, has attempted to bypass all these difficulties. If the good for an individual decision-maker is that which is judged to maximize expected utility, then there is no strong demand for getting the causal story correct about how much utility will be produced. It is merely a question of what the agent expects to occur. Also, if all the evidence for an agent's evaluation is an expressed preference, then the distinction between quantity and quality drops out along with the demand to measure the subjective happiness states of others. Reducing cognitive load down to this level, however, raises the question of why preference utilitarianism should count as a moral theory at all. Why is it not merely a description of an agent's expectations and preferences? The answer, briefly, is that the descriptive task of explaining an agent's moral choice-making in terms of maximizing expected utility does double duty as a theory of rationality, with the suppressed normative premise being that one ought to act rationally.

There are at least two significant issues raised by this version of preference utilitarianism. The first is that equating normativeness with the demand to be rational is regarded by some as committing the naturalistic fallacy, the supposition that it is a fallacy to equate a moral property (e.g. goodness, rightness, justice) with a natural property (e.g. happiness, growth of knowledge, rationality). Because of our semantic holism, we are not much impressed by this concern, though it does present one issue that we shall touch on later. The second we regard as rather more serious. Once rationality is thought to have normative force, it needs to be more than a matter of just having a consistent preference structure. The expectations that feed into the construction of utility functions as assigned probabilities — our estimates of the likelihood of expected events occurring — need to be warranted. Without warrant, ignorance compromises the assumption of rationality. However, meeting this demand now reintroduces the same cognitive load problem about computing causal consequences of actions performed in complex social contexts as that which attended classical utilitarianism.

Not surprisingly, the demands of moral leadership require a certain amount of cognitive elitism, drawing on skills of situation analysis, a grasp of the causal workings of complex social scenes, a well structured set of preferences, and a knack for calculation. Given that a technical result due to Kenneth Arrow means that there is no rational way for aggregating individual judgements of utility into a function that maximizes collective well being, elitism, on this view, is essential since someone's preferences must prevail in the collective. (See Arrow 1963; Evers and Lakomski 1996, pp. 154–164.)

The other major relevant tradition in moral reasoning, after varieties of utilitarianism, is Kantianism. Broadly speaking, on this approach particular moral precepts are evaluated in the light of some general canon, or canons, of rationality. Perhaps the best known modern example is John Rawls's (1971) attempt to demonstrate what principles of justice would be chosen by persons acting rationally under conditions of impartiality. The argument is based around a thought experiment where people, unaware of what position they would occupy in a society, reason about what principles of justice should regulate their social life. We don't want to go into any detail about the theory, but we do want to indicate a consequence for practical application when it comes to moral guidance. Once impartiality is construed as requiring ignorance of organizational or social detail, the deduction of principles becomes a very abstract and intellectually demanding task. Moreover, we end up with linguistically expressed principles of a fairly high level of generality, sufficiently so to cause a version of the generality problem afflicting moral codes to break out. But once we attempt to ease the cognitive burden by plugging in familiar social detail, the impartiality that is a presumed condition of valid reasoning about circumstances that includes oneself, is compromised. The resulting Kantian moral leader is stereotypically familiar: highly rational in defence of principles, and affecting a detachment irrespective of the pattern of disbursement of their material consequences.

In dealing with ethical guidance from theories of moral reasoning, the worry is not primarily about a lapse in coherence between the ethical component and the theory of leadership required to implement it. The worry is that the demands of implementation expose weaknesses in the psychological plausibility of both bodies of theory. For unfortunately, as we have been arguing throughout, several serious problems attend the predominantly linguistic-computational view of cognition that underwrites much thinking about moral knowledge. First, the organ that does the computation, for example, adjudicating the relative probabilities of alternative expectations, attaching weighted preferences to each and multiplying out the matrix of results, is not a computer but a brain whose processes of decision-making are known to be quite different in operation. (See Evers 1998, for an overview.) Second, in most cases of decision-making, or what might be classified as intelligent action, there are no symbolic structures at all to operate on. For the countless acts of judgement that are performed every day, people classify, sort, prioritize, adjudicate, and recommend without the benefit of any language-like theory formulation of the issues at hand. Much of this cognitive activity is better seen as a species of pattern processing; for example, of processing visual representations of complex scenes associated with work or interpersonal matters. Indeed, we claim that most knowledge to do with skilled practice does not exist in the form of language-like representations. Third, although computational models of cognition devote attention to the sequential ordering of processes, they give no detailed consideration to real time processing and its consequences for decision. The psychology of deliberation is replaced with the logic of calculation. Unfortunately, the replacement is not without loss. For example, we know that deliberation time affects decision outcomes. We also know that non-equivalent descriptions of the same outcomes can affect rankings of values. Fourth, computationalism's focus on logical and quasi-logical relations among symbolic representations fails to mesh with the most developed accounts of cognition in terms of the causal machinery of human learning and information processing. But for computationalism to be an account of human cognition some rapprochement is called for. Finally, computational models lack links with epistemology, with accounts of how knowledge is built up. Yet without these links, their capacity to offer justifications of posited expectations is compromised

Meeting these difficulties, for practical knowledge in general, but also for ethical knowledge in particular requires, in our view, an account of knowledge representation that coheres with a naturalistic view of cognition as neural information processing. Let us review the main features of our approach, for which the primary elements of cognition are non-symbolic, in terms of the sort of practical knowledge assumed for leadership.

Take the case of a school principal who has to make regular judgements about admitting children with special needs into a mainstream school environment. In common with most practical problems, it is not possible to formulate a decision rule of the form: 'admit student $X$ when conditions $C$ obtain'. The issue is really a cluster of ethical, social, interpersonal, resource, and policy considerations with a smooth continuum of relevant factors falling between a decision to admit or to refuse. No symbolic theory formulation is ever likely to be helpful in deriving a conclusion and none is ever used in real time practice. Yet after what are agreed to have been a series of decisions of mixed success, learning has occurred, with subsequent admission decisions being regarded as mostly right, or appropriate.

Schematically, the learning framework we are suggesting includes the following elements. A person brings to experience some prior set of dispositions that provide an initial set of classifications, or similarities, and saliences. These serve to group the passing show into kinds that elicit responses, in turn subject to interpreted feedback that leads to a change in our initial stock of knowledge. We learn to recognize socially important kinds, for example, people who are helpful, or competent, or even good. In organizational settings we might classify staff into those easy to work with, innovative, task oriented, or difficult. And we would include in the way we extract patterns from experience, observable clues to the existence of these social kinds, so that we might more efficiently recognize them in the future. Learning occurs over time when the accumulating knowledge, or map by which we steer or navigate our way around the world, results in an increasing, non-random, chance of success relative to particular tasks.

Using the model of learning by backpropagation of error correction in artificial neural networks, a further level of abstraction for this example is possible. Thus, an initial typology of relevant features, for example, type of impairment, level of support required, available resources, distance from school, availability of alternatives, etc. would function as the input nodes, with some judgement of degree of presence serving as the values making up the input vector. The target output could be just a two-value vector indicating either a satisfactory or unsatisfactory outcome. Multiple experiences, where the principal learns what degree of features are associated with successful and unsuccessful outcomes produce an appropriate set of weights for classifying input vectors into two prototypes for admission — those features of admitted students that make for successful integration and those that do not.

There is no expectation that the graded nature of classification by similarity to learned prototype, easily represented by the non-linear mathematics of the net and its operation, can be captured by a linguistically expressed formulation. The principal's competence is not to be found in any linguistically formulated decision rule that the principal is able to express. From this portrayal of learning from experience there is no such rule. Rather, competence is a matter of efficient learning and resulting good judgement in the contexts at hand. There is no essence, or even a lawlike generalization about success in admissions practice and this is true for the vast bulk of practical judgements and decision tasks that are made every day. Generalizing a bit, we can make the same point about the constellation of dispositions that make up a practical skill like leadership. Exhibited under many circumstances, with many outcomes, and subject to the interpretations and behaviors of many participants, the conditions under which we would learn to identify and classify acts as instances of leadership are most likely to reflect all the features associated with learning from experience. Through a process of mutual adjustment of expectation and performance perhaps funded initially from everyday commonsense conceptions embedded in the learning of language, we build up a leadership prototype, though one located within the circumstances of experience. Under these conditions, the quest for even a set of common characteristics, is most likely a forlorn quest, as reasonable as supposing in advance of inquiry that Gandhi, Thatcher, Stalin, or Boutros Ghali must have some special feature in common. It is simply an open question whether the prototypes developed through experience in one context are useful in other contexts — a matter for further experience.

As an interesting corollary to the above reflections, note that on this view of practical knowledge, many of the generic functional concepts of administrative theory outrun the available cognitive evidence for their successful implementation. So leadership that is expected to produce large scale social change contrary to the normal operation of major institutions will probably require a prototype of power instantiated in the minds of a followership that lies well and truly beyond the local and particular orbit of experience of successful piecemeal change. Experience is more able to fund a practical notion of effective action than a generic notion of power. (Robinson 1994) Of course, generic theories of power do exist in symbolic form, but these are often accessible only to elites, with the usual political risks of vanguardism and the social production of new hierarchies.

## Representing Ethical Knowledge in Leadership

When it comes to ethical knowledge, the chief point we want to make is that it is acquired in the same way as other practical knowledge — mainly through learning from experience in complex, shifting, context bound circumstances. That is, categories of moral appraisal are prototypical patterns extracted from experience through the epistemic practice of learning. To develop an account of ethical knowledge that coheres with a naturalistic view of moral agents, and which counts ethics as on a par with any other knowledge that is acquired and validated through the usual processes of learning, we want to return and address a still lingering issue concerning the so-called 'naturalistic fallacy'.

Naturalism in ethics involves two main components. (Flanagan 1996, pp. 193–194). The first is a descriptive-genealogical component, concerned with describing our moral dispositions, their origins, and their operation in deliberation. Moral psychology is part of this domain, as are sociology and history. There is little controversial about the project of a naturalized ethics in this first sense. There is, however, deep controversy about naturalizing ethics in a normative sense. For here additional claims need to be made and defended, not about further information on how we arrived at the moral judgements we make, but whether these moral judgements are good or bad — whether they are normatively appropriate. That is, when all the facts are in concerning the totality of our moral behaviors, when the descriptive-genealogical story has been completely told, there still remains the question of whether these behaviors are ethical. However, the challenge is not to produce a definition of 'ethical' that is analytic. Our view of semantics makes this unnecessary. Rather, the task is to locate any purported norms within a more coherent framework. Not surprisingly, the candidate we favour links the development of ethical knowledge to what we take to be a set of defensible epistemic practices — practices that over the medium to long run do better than chance in leading to reliable knowledge. The naturalistic philosopher Paul Churchland (1989, pp. 301–2) deals with this issue in a particularly insightful way:

> When such powerful learning networks as humans are confronted with the problem of how best to perceive the social world, and how best to conduct one's affairs within it, we have equally good reason to expect that the learning process will show an integrity comparable to that shown on other learning tasks, and will produce cognitive achievements as robust as those produced anywhere else. This expectation will be especially apt if, as in the case of 'scientific' knowledge, the learning process is collective and the results are transmitted from generation to generation. In that case we have a continuing society under constant pressure to refine its categories of social and moral perception, and to modify its typical responses and expectations.

What Churchland has in mind as an account of social and moral learning is the neural network story, with children building up their social and moral categories from infancy onwards, through a naturalistic process of coherently matching feedforward expectations against feedback from outcomes to produce a socially useful fit. The reason this process can be regarded as normative rather than merely descriptive is because it involves a critical dimension that makes it more than mere socialization. (See also Churchland 1995, pp. 123–150.)

In Owen Flanagan's (1996, pp. 206–207) terms...

> Social experience provides feedback about how we are doing, and rational mechanisms come into play in evaluating and assessing this feedback. So there is an aim, activity to achieve this aim, feedback about success in achieving the aim, and rational mechanisms designed to assess the meaning of the feedback and to make modifications accordingly.

This puts the case for ethical naturalism squarely where it belongs: in the company of a naturalistic tradition that links the growth of ethical knowledge with a defence of the possibility of good epistemic practice for all knowledge. And on the view of natural knowledge representation as a geometric configuration of distributed weights, the quest for universal moral rules, or context free moral generalizations, is misguided, more an artifact of linguistic representations than a reflection of the myriad of details that go into the critical development of a learned moral prototype.

From the vantage point of the new cognitive science, the link between leadership and ethics in administrative contexts can be characterized roughly as follows. It is doubtful if the successful solution of diverse human problems can be explained by their possessing some essence, or even some feature that they all have in common. 'Advancing human flourishing' is the usual formulation, but it is as normatively useful as a guide to practice as 'doing good and avoiding evil' is. These are linguistic formulations which take their place in the construction of rules where in fact little or no rule following occurs. People merely behave, most of the time, *as if* they are following rules. The conditions circumscribing the development of ethical prototypes are diffuse and fragmented. However, a condition of their acceptability is that they are the product of progressive epistemic practices — practices that reduce the randomness of our response to the passing show of experience and increase the amount of information in the coherent global map by which we navigate our way through the option spaces of social life. Moral leadership in the contexts of organizational life is therefore a matter of securing the social conditions of effective learning in these contexts. As Dewey and others have seen, the conditions for the growth of knowledge involve an ethical infrastructure. That is, knowledge, including moral knowledge, develops more efficiently under some ethical arrangements than others. Nor are these arrangements surprising or novel. The progressive application of feedback against bias and error requires freedom of speech, tolerance of opinion, and respect for persons and their right to participate in the growth of knowledge. There is also a host of more derivative imperatives, expressible in the broad-brush strokes of language.

This point at which the theory of leadership and the theory of ethics converge, namely over the question of how to solve the epistemic problem, or how best to arrange matters so as to navigate successfully the complexities and uncertainties of social and organizational life, suggests useful possibilities for the theoretical development of each. Earlier, we claimed that people may build up prototypes of leadership through the sifting of examples in shifting contexts. However, where leadership involves an ethical dimension, an unrealistic cognitive load can accrue to the leader under some of these prototypes. Ironically, the usual way in which cognitive load is diminished for individual cognizers is by the adoption of organizational structures that enhance learning through the processes of distributed cognition. (Lakomski and Evers 1999.) In keeping with this strategy, the acquisition of moral knowledge as well as much other knowledge relevant to decision-making within the complex uncertainties of organizations, will benefit from the adoption of a more distributed model of leadership committed to organizational learning. The ethics thus acquired can be applied recursively to the problem of what ethical arrangements among people in social life can best provide for more ethical and other learning. In this way, a view of ethical leadership can also have moral value.

While the above characterization of ethics is fairly general, even though it has been located more particularly within the constraints of leadership theory, it does lend itself to suggestions on how to proceed on ethical issues. It is useful to see these suggestions as taking the form of a sequence of steps:

(a) The first step in ethical judgement and action, which is analogous to the well-known task of values clarification, involves trying to identify your own prototypes associated with each moral issue in question; for example, what you think are the main features associated with being fair, or good, or acting appropriately.

(b) Then, look at the features of the *context* in which your action or decision is required, doing what amounts to situation analysis concerning, for example, your particular school, its environment, staffing and student profile, budgetary considerations, etc.

(c) Next, examine how your ethical prototypes apply in that context, whether, say, fairness is instantiated, or rightness. This examination functions as a *critical application* from which learning should occur, notably whether your identified prototypical values are adequate for dealing with the situation *and* whether the context or situation can be improved or changed in accord with your values.

(d) At this point, some adjustment to the identified set of prototypes comprising aspects of your values framework, may be required in order for its elements to cohere with values implicit in the exercise of critical learning.

(e) In making an ethical response in some circumstance, the notion of prototypical contexts is relevant. If your are a leader, then one valuational constraint will be the requirement to act in a manner true to expectations arising out of your leadership prototype. If the context is a school, an understanding of schools will provide further overlay on the notion of leadership involved. And if a school has particular characteristics, as all schools do, still more refinements will need to be made.

(f) One point at which central normative features of particular organizational contexts emerge and are given expression is in vision and mission statements, or in schools, the equivalent in the form of school charters, school plans, and the like. To the extent that these circumscribe options for action, the resulting choices will reflect a process of mutual coherent adjustment between your values framework on the one hand and your (partly) normatively driven understandings of roles, organizations and the particularities of contexts on the other.

(g) Finally, the consequences of ethical judgement and action constitute the data of ongoing critical learning by which we attend to the education of our global web of belief. In this way, improved ethical knowledge and practice emerges naturally out of learning to navigate our way through the problems and issues of everyday organizational life.

The extent to which the above process schema is generalizable is not so much problematical as in need of further specification. For example, outside of leadership roles, the scope for learning will need to be unpacked. Similarly, accounts of what is reasonable to expect of a person acting in some other organizational role will need to be given. And so on. However, where these specifications are made, the normative goal of them becoming grist for the critical learning mill does not lose its force.

In the next chapter we deal more systematically with the question of administrator training, taking the opportunity to place in a wider perspective the issue of training for critical learning.

# References

Arrow K. (1963). *Social Choice and Individual Values*. (New York: Wiley).

Ayer A.J. (1946). *Language, Truth and Logic*. (New York: Dover).

Churchland P.M. (1989). *A Neurocomputational Perspective*. (Cambridge, MA: MIT Press).

Churchland P.M. (1995). *The Engine of Reason, the Seat of the Soul*. (Cambridge, MA: MIT Press).

Churchland P.M. (1996). The neural representation of the social world, in L. May, M. Friedman, and A. Clark (eds.) *Mind and Morals*. (Cambridge, MA: MIT Press).

Churchland P.M. (1998). Towards a cognitive neurobiology of the moral virtues, *Topoi*, **17**, pp. 83–96.

Evers C.W. (1998). Decision-making, models of mind, and the new cognitive science, *Journal of School Leadership*, **8** (2), pp. 94–108.

Evers C.W. and Lakomski, G. (1991). *Knowing Educational Administration*. (Oxford: Pergamon Press).

Evers C.W. and Lakomski, G. (1996). *Exploring Educational Administration*. (Oxford: Pergamon Press).

Flanagan O. (1996). The moral network, in R. N. McCauley (ed.) *The Churchlands and Their Critics*. (Oxford: Blackwell).

Foster W. (1985). *Paradigms and Promises*. (New York: Prometheus Books).

Greenfield T.B. and Ribbins, P (eds.) (1993). *Greenfield on Educational Administration: Towards a Humane Science*. (London: Routledge).

Habermas J. (1972). *Knowledge and Human Interests*. (London: Heinemann).

Hodgkinson C. (1991). *Educational Leadership: The Moral Art*. (Albany, NY: SUNY Press).

Lakomski G. and Evers C.W. (1999). Values, socially distributed cognition, and organizational practice, in P. Begley (ed.) *Values and Educational Leadership*. (Albany, NY: SUNY Press).

Leithwood K., Tomlinson, D. and Genge, M. (1996). Transformational school leadership, in K. Leithwood *et al.* (eds.) *International Handbook of Educational Leadership and Administration*. (Dordrecht: Kluwer).

Newell A. and Simon, H.A. (1976). Computer science as empirical enquiry: symbols and search. Cited as reprinted in Boden, M. A. (ed.) *The Philosophy of Artificial Intelligence*. (Oxford: Oxford University Press).

Popper K.R. (1945). *The Open Society and its Enemies*, Volumes I and II. (London: George Routledge and Sons).

Quine W.V. (1951). Two dogmas of empiricism, *Philosophical Review*, **60**, pp. 20–43.

Quine W.V. (1960). *Word and Object*. (Cambridge, MA: MIT Press).

Rawls J. (1971). *A Theory of Justice*. (Cambridge, MA: Harvard University Press).

Robinson V.M.J. (1994). The practical promise of critical research in educational administration, *Educational Administration Quarterly*, **30** (1), pp. 56–76.

Simon H.A. (1946). *Administrative Behavior*. (New York: Free Press).

Simon H.A. (1976). *Administrative Behavior*. (New York: Free Press, 3rd Edition).

Small R. (1998). Towards an unprincipled ethics of educational research, *Australian Journal of Education*, **42** (1), pp. 103–115.

Smith M. (ed.) (1995). *Metaethics*. (Aldershot, UK: Dartmouth).

Wong K.C. (1996). Preparing educational leaders in moral education: a reflection on traditional Chinese leadership thinking. Paper presented to *APEC Educational Leadership Centres*, Chiang Mai, Thailand.

# 8

# Administrator Training

Herbert Simon, while presumably not thinking about the training of administrators especially, nevertheless formulates the problem to be discussed here in a manner suitable for our purposes:

> A central design issue, when we are planning learning experiences for our students ... is how much and what kinds of transfer of knowledge we can expect from the specific content of textbooks, lectures, and homework problems to the tasks that students will be expected to handle in subsequent courses and in professional life (Simon 1980, p. 81).

Simon (1980, p. 82) also acknowledges that the empirical evidence for transfer is mixed, a finding confirmed by more recent research (Anderson, Reder, and Simon 1996; see Chapter 3, Section 4). Indeed, Sternberg and Frensch (1993, p. 25) comment that 'transfer of training often appears to be the exception rather than the rule, whether in school or outside of it.' This state of affairs is said to be characteristic of much of the preparation educational administrators undergo. Programs taught — at college or university — are said not to have contributed in any significant degree to making more competent professionals. The *problem of transfer* which, when reframed, amounts to the problem of how to create expert administrators, is thus a key concern of administrator preparation programs.

In this chapter we examine some proposals representative of recent reforms suggested by Murphy (1992), Bridges and Hallinger (1995), and in particular the *cognitive apprenticeship (-in-problem-solving)* model advocated by Prestine and LeGrand (1991) and Prestine (1993) which take issue with traditional, university-based, didactic programs. The latter are believed to be either limited or unsuitable for use in professional practice because the abstract knowledge taught is allegedly too removed from professional practice which is characterized by the need to solve ill-structured problems. Since these problems change constantly, a canon of abstract knowledge, learnt in situations quite unlike those of practice, is unlikely to be useful. The consensus among contemporary developers of preparation programs, even though there are differences of detail, is the shared belief that good administrators acquire their expert professional knowledge in practice through years of personal experience, aided more or less by formal studies. In this way, they appear to believe, the *problem of transfer* is avoided or simply does not arise. If this assumption were correct, then there is indeed hope that the training of administrators would lead to improved practice and create professional expertise.

In our view, these recent moves are to be supported in their general orientation since they stress the importance of situation and context in the acquisition of professional competence. As is evident throughout our treatment of naturalistic practices in this book, we often have cause for agreement with current strategies and proposals for practice as discussed in the literature. Nor should it surprise that such agreement exists. We differ, however, in terms of how to explain the functioning and success (or otherwise) of such proposals. To the extent that administrators have been able to learn how to be administrators, and become competent in their fields, the various experiences they have encountered must have been successful learning (or transfer) experiences of some kind — whether they were encountered in formal university classrooms (through the transmission mode), by shadowing administrators, or by actually doing the job (through situated experience). Transfer, however defined, must have taken place.

We argue here that contemporary preparation reform programs, and thus the promise of creating improved practice, will be strengthened once some confusions have been eliminated. The central issue is the all pervading distinction between *active* (situated; practical) and *passive* (university-based; abstract) knowledge which hides a mistaken view of how humans learn and what the nature of human cognition consists of; this mistaken *symbol processing* view equates all human cognition with what can be *represented symbolically*. It does not, however, represent the administrator's practical knowledge since it is not of a symbolic form. On this *mistaken* account, then, even if there were positive transfer of learning we would not know about it because competent practice is inaccessible to symbolic representation. Considered from the reform perspective, belief in the superiority of *situated* learning or cognition over 'transmitted' abstract knowledge is equally mistaken.

The important point to be made is twofold and has considerable implications for the training of administrators. Practical knowledge is not practical because it is locally produced, the assumption made by the reformers including situated cognitivists; it is practical because it is *sub-symbolic*. By the same token, textbook and other forms of formal knowledge is not abstract because taught in formal settings; it is abstract in that symbolic representations such as language are collective externalized tokens of accumulated and compressed social experience, past and present. It is the confusion between assumptions of how and where we learn, and how this knowledge is represented which causes the problem, in traditional as well as current reform programs.

Seeing human beings as pattern recognizers dissolves the distinction between active and passive, practical and theoretical knowledge, and makes possible a more productive assessment of symbol processing since symbols *are* patterns. In addition, since humans are pattern recognizers *in social contexts* such as schools and universities, for example, we expand on the importance of context or situatedness from our naturalistic perspective. We also add some recent insights from connectionist cognitive science (Churchland 1998) which explain, for the first time, how it is possible for our individual and idiosyncratic brains to detect conceptual similarity. An account of this kind has been missing in neural net research; it is also the kind of account needed by situated learning theorists to explain how people are able to transfer practical knowledge learnt in one setting to another setting.

In the following, we begin by considering some of the contexts in which the transfer problem has been addressed by examining how administration theorists thought they could improve the training of administrators in order to create competent professionals and thus improve education practice.

## The Science/Craft Distinction in Administrator Preparation Programs and the Problem of Rational Administration

Administrator training, just as the business of administration in general, has traditionally been viewed as either a scientific/technical or an art/craft-type activity. Dealing with administrative practice in the former view is seen as a matter of rational problem-solving and decision-making, as prominently suggested by Simon (e.g. Simon 1945, 1977, 1980). Hence, the best administrator preparation consisted of teaching the general principles of problem-solving in the hope of providing the most rational means to solve the problems of practice. Ernst and Newell's (1969) General Problem Solver (GPS) provided an influential early computer model which offered a range of general problem solving techniques applicable to very different problems. Its characteristic features were that it employed a means-ends model; the problems tackled were well defined in terms of initial and end states, as were the operators employed. This model has without doubt been the most influential in administrative theorizing and has influenced administrator preparation in various forms until the present. (For more detail on the empiricist model of administrative theory and practice see Chapter 3, Evers and Lakomski 1991).

At the other extreme, Wagner (1993, pp. 88–89) notes, are those researchers 'who view management as a craft [and who] are more likely to study experienced administrators rather than computer programs.' Greenfield, for example, is one writer who repeatedly emphasized the complexity of administrative action and problem-solving which was not likely to be solvable by rational means. Rather, on the issue of administrator preparation, he advocated 'placing the novitiate in an actual monastery ….They might spend time as a bartender, bouncer, or manager of a disco, followed by service as an orderly in a mental institution, or indeed as a patient … '. (Greenfield and Ribbins 1993, p. 112). Becoming an effective administrator, then, is in his view not a matter of specific training programs organized around rational principles but a matter of diverse subjective experiences.

Considering the rational model of administrator preparation, the idea that expertise consists in the application of general theory or principles has been subjected to criticism by Kennedy (1987) in her detailed treatment of professional education and the development of expertise. She points out, amongst other things, that there are three questions regarding the presumption that general principles and theory can be applied to specific situations:

> The first question has to do with how the practitioner recognizes a particular case as an example of a general principle; the second with how the practitioner adjusts predictions derived from a general principle to accommodate the special features of the case; and the third with how practitioners blend the variety of potentially relevant principles to form an integrated body of knowledge that can be applied to specific cases. (Kennedy 1987, pp.139–140)

As she rightly notes, real cases do not appear in form of general principles and people have to find out first what the relevant features are which tend to be embedded in much complex detail. 'Thus', she says (Kennedy 1987, p. 140), 'expertise is not merely the knowledge that general principles exist; it is the ability to recognize the cases to which they apply'. With regard to the second question, Kennedy argues that since most principles which are derived scientifically are probabilistic, 'even a correct diagnosis of the case as an example of a principle may not indicate that the principle should be applied to that particular case'. The third question asks how practitioners actually select between competing principles, a difficulty she neatly sums up as follows: ' … principles provide rules of thumb intended to guide practice, but there are no rules of thumb for how to select the appropriate rule of thumb' (Kennedy 1987, p. 142). In addition to the difficulties of when to apply principles, there are others which relate to whether human beings can actually make such rational decisions. It appears that they cannot.

We have known for some time that even the most effective managers do not manage according to rational procedures and means (e.g. Mintzberg 1973; Mintzberg, Raisinghani, and Théoret 1976; Argyris and Schön 1996). In addition, real-world problems, because of their complexity, diffuseness and intractability, simply evade the grasp of rational techniques designed for generality of approach and outcome. Furthermore, even if managers and administrators were to proceed in the rational problem-solving mode, we know from research in social psychology that human reasoning and judging is prone to biases which affect human problem solving (Nisbett and Ross 1980; Tversky and Kahneman 1983, 1986) including acquisition, processing, and response biases (for more detail see Wagner 1993, pp. 92–94).

Discontent with the traditional scientific model of administrator preparation is certainly not the prerogative of the current reformers. Twenty-five years ago, March (1974) famously took stock of the state of affairs of university-based administrator preparation programs. He issued a warning that one ought to be cautious in attempting to identify generic needs for educational administration which is too diverse a field, and in dealing with educational institutions which are 'organized anarchies' made up of problematic goals, unclear technologies and fluid participation (March 1974, p. 20). More important still was his reference to the 'ideology of administration' (March 1974, p. 18) which has become famous in the field:

We can characterize that ideology by the following set of beliefs. If there is a problem there is a solution. If there is a solution it can be discovered by analysis, and implemented by skill in interpersonal relations or organizational design. The solution to a problem requires the identification of underlying causes and the discovery and implementation of solutions are duties of the administrator. If a problem persists, it is due to the inadequacy in an administrator's will, perception of problems, analysis, skill with people, or knowledge of organizations. Inadequacies in an administrator can be corrected through proper administrative training.

This 'faith of hope', March (1974, pp. 18 onwards) observed, is ill-founded. Problems may not be amenable to solution due to complexity; we may discover solutions by means other than analysis; conflict may not be resolvable and complex implementations may outrun a solution's complexity or the capabilities of participants; removing causes may be the least likely way to solve a problem. Problems may simply persist and be in principle beyond the scope of the administrator's influence. He concludes that we should not expect that a change in administrator training can eradicate the problems of modern education. And yet, he does not abandon the field but rather suggests that administration as 'the art of intelligent coping with an arbitrary fate' (March 1974, p. 23), while being a minor matter, may still be helpful since 'our prospects for human control over events are built on collections of minor matters.' March's insights and warning remain valid even in light of new proposals for changing administrator preparation programs (some of) which acknowledge the limitations of human influence and control over external events. We have occasion later to comment on the over-reliance on the cognitive capacities of the individual administrator so clearly flagged by March.

## Practice-based Administrator Preparation

It is not our aim to provide a comprehensive examination of contemporary reform proposals in the present context. Rather, we want to focus on some recent and prominent examples in order to show the direction of change which continues to take place in the field, and consider the arguments advanced in favour of practice-based programs.

In *The Landscape of Leadership Preparation*, Murphy (1992, p. 140), for example, suggests that the direction of training has been inappropriate. Rather than beginning with prescriptive bodies of knowledge for administrator training, which had been the case for the first fifty years of the field, we should 'backward map from the goals of preparation, rather than vice versa' and begin with better defined program goals. In his view, ethical and intellectual concerns rather than administrative roles should be the main platforms for administrator preparation. More specifically, administrators should be prepared to become 'moral agents, educators, inquirers, and students of the human condition.' The first goal leads to de-emphasizing administration as a science and focuses on the character of the administrator as a moral agent; the second goal serves to remind administrators that the business of administration is part of the larger field of education rather than generic administration; teaching inquiry skills is more important than assuming that there will be transfer of knowledge since, in Murphy's view, it has not been possible to establish a 'codifiable knowledge base' in the history of administration and training. It is much more important to emphasize the construction of knowledge and de-emphasize its dissemination, indicating a 'paradigmatic shift' to 'constructivist approaches to learning' (Murphy 1992, p. 145). The final goal Murphy identifies, the human condition, follows on from the earlier ones in that it stresses that administrators are broadly leaders of human beings, and that people should be treated as ends rather than means.

The principles which underlie Murphy's proposals are broad in character, comprise new prescriptions for curriculum design, and include explicit rejection of the following still operating norms:

> the belief that administrators can be prepared to deal with the specific content of their jobs, and that we can do this better by preparing people for ever more discrete roles; (b) equating preparation with the transmission of a 'systematized body of knowledge' … — either discrete technical skills or discipline-based content; (c) the separation of administration from education and values; and (d) distinctions between theory and practice (Murphy 1992, p. 147).

It is clear from his description that his proposals differ substantially from the rational problem-solving approach favored by Simon, for example. Murphy is particularly concerned to emphasize that administrator training, on the basis of the principles mentioned above, can now be seen as 'situating learning in context', leading to a 'learning-in-action' context for trainee administrators, a more *problem-based approach*. A number of practical proposals for the structure and content of preparation programs are suggested which reflect these views.

Broadly sympathetic to Murphy's orientation, Bridges and Hallinger (1995) offer an approach to the training of administrators which focuses directly on problem-based learning. (This approach is not to be confused with the broader methodological approach developed by Robinson. Her proposal for problem-based learning is more comprehensive and while not directed towards administration especially, it also applies to the training of administrators. It addresses especially situational constraints which shape problem definition *and* solution. See Robinson 1993, 1998)

Bridges and Hallinger are quite as adamant as Murphy that their proposal differs sharply from traditional programs. In particular, they disagree with the following four assumptions which, they claim, underlie traditional preparation programs:

> (1) the knowledge is relevant to the student's future professional role; (2) learners will be able to recognize when it is appropriate to use their newly acquired knowledge; (3) application of this knowledge is relatively simple and straightforward; and (4) the context in which knowledge is learned has little or no bearing on subsequent recall or use. (Bridges and Hallinger 1995, p. 5).

Problem-based learning, on the other hand, relies on different assumptions. The main emphasis is on *knowing* and *doing*. Students are supposed to learn better when their prior knowledge is taken into account and when they are capable of making links between new and old knowledge. They are encouraged to apply their new knowledge, and do so in situations which resemble as much as possible the actual contexts of future use. In addition, Bridges and Hallinger believe that it is important to employ problems students will be likely to encounter later in their professional lives in order to provide a meaningful context for application. Their model for professional development is based on five interrelated issues which complement the proposal offered by Murphy. Of central importance are (1) the workplace, (2) goals, (3) content, (4) process of teaching-learning, and (5) student evaluation. With regard to four, the authors consider their model to be radically different from traditional approaches in that the *project* is based at the centre of the instructional process. This is a kind of simulation which, like real life problems in administration, is ill-structured. As such, problem based learning is close to what is called the case method in administrator preparation and appears to differ more in nomenclature than substance although Bridges and Hallinger mention some differences in the means of instruction and student evaluation.

Considering the two positions of administrator preparation considered so far, it would appear that neither Murphy nor Bridges and Hallinger believe that there is much benefit in the teaching of a college or university based body of knowledge because of lack of subsequent transfer. Both positions rather advocate the construction of administrative knowledge in the context of a simulation or in connection with a problem likely to be encountered in the professional lives of administrators. The role of formal taught knowledge remains somewhat peripheral.

The main beliefs underlying these proposals is the assumption that administrative competence and expertise is acquired by practicing administration rather than by being instructed on how to do it. While both prominent approaches stress context and activity as fundamental, others address administrator preparation from an expressly cognitive perspective and thus introduce a focus on learning rather than teaching, or being taught. Leithwood, Begley and Cousins (1994), for example, argue for an information processing view and advance a theory of learning which is based on social interaction and adult learning theory. The conception, however, which focuses most directly on both *situated* practice and how administrators learn their craft, is the notion of *cognitive apprenticeship*, (see Collins, Brown and Newman's 1989 original definition) as developed by Prestine and LeGrand (1991) for educational administration, and extended by Prestine (1993) in her concept of *apprenticeship-in-problem-solving*.

## Administrator Preparation as Cognitive Apprenticeship

Since administrative practice is characterized by having to understand and provide solutions to ill-structured problems, it is not possible to determine and subsequently draw on a body of knowledge in advance which can be expected to transfer across various settings and problem situations. Indeed, Prestine (1993, p. 204) is convinced that 'most educational administration preparation programs are designed around a profoundly misleading idea: that one first acquires administrative knowledge and later applies it in practice.' Instead, she argues (Prestine 1993, p. 193), we need to 'focus on *how* knowledge is used in practice, rather than on *what* knowledge structures are.' The goal of administrator preparation programs then becomes 'the acquisition of generative knowledge with wide application in novel but partially related contexts … [and] to promote knowledge acquisition for later accessibility and use in practitioner problem-solving situations' (Prestine 1993, p. 208). Prestine observes that expert practice to date, to the extent that an administrator has been able to develop it at all, has been the result of accumulated field experience. The model of *cognitive apprenticeship-in-problem-solving*, she argues, by becoming an integral part of any preparation program, is designed to take administrative expertise 'out of the realm of haphazard acquisition.' Programs thus redesigned would remove 'the dominant role of isolated, passive, and sterile knowledge acquisition as the primary activity of preparation programs' (Prestine 1993, p. 208).

The notion of *cognitive apprenticeship*, as originally developed by Collins and colleagues (1989), derives from cognitive learning theory which advocates situated cognition as a concept able to explain how novices acquire the relevant *practical knowledge* which in principle leads to competence and expertise. Primarily designed to explain how children learn to read, write and do mathematics, Collins *et al.* (1989, p. 459) differentiate their model from traditional apprenticeship 'in that the tasks and problems are chosen to illustrate the power of certain techniques or methods, to give students practice in applying these methods in diverse settings, and to increase the complexity of tasks slowly … tasks are sequenced to reflect the changing demands of learning'. Secondly, they are keen to stress that, unlike traditional apprenticeship models, they wish to *decontextualize* knowledge, by which they mean that knowledge can be used in many and diverse settings. The writers also wish to ensure that teachers explain abstract principles underlying the application of knowledge and skills as experienced in the different situations. The specific contribution their model is said to make is located in its 'dual focus on expert processes and situated learning-through-guided-experience on cognitive and metacognitive, rather than physical, skills and processes' (Collins *et al.* 1989, p. 457). This is how Collins *et al.* expect to be able to overcome the 'educational problems of brittle skills and inert knowledge'.

For Prestine and LeGrand this model offers a powerful conceptual tool for tackling the reality of the practitioners' world in administrative contexts and also facilitates the reconsideration of related areas such as Schön's reflective practice, the concept of novice versus expert practice, as well as enabling renewed emphasis on ethical and moral considerations (Prestine and LeGrand 1991, pp. 62–63). Of particular significance for a new model of administrator preparation is situated cognition's assumption *that culture and cognition are inextricably linked*. As they point out, '… knowledge is both incorporated into the learner's existing repertoire of knowledge and made meaningful for the learner by the context and activities through which it is acquired' (Prestine and LeGrand 1991, p. 63). Unfortunately, they continue, 'professional preparation programs and views of professional knowledge creation and acquisition have either ignored or scorned such precepts.'

Following Collins *et al.* (1989), *cognitive apprenticeship* consists of four dimensions which together make up an 'ideal learning environment': content, method, sequence and sociology. By content is meant the distinction between conceptual theoretical knowledge, traditionally taught in academic administrator preparation courses, and the tacit or *strategic* knowledge applied in actual problem-solving and incorporating reflection on action and strategies employed in learning new concepts and facts. The teaching methods deemed more appropriate fall into three groups:

> ... the first three (modeling, coaching, and scaffolding) are the core of cognitive apprenticeship, designed to help students acquire an integrated set of cognitive and metacognitive skills through processes of observations and of guided and supported practice. The next two (articulation and reflection) are methods designed to help students both focus their observations of expert problem solving and gain conscious access to (and control of) their own problem-solving strategies. The final method (exploration) is aimed at encouraging learner autonomy, not only in carrying out expert problem-solving processes, but also in defining or formulating the problems to be solved. (Collins *et al.* 1989, p. 481).

Sequencing learning or knowledge acquisition is important in that it should strive to increase both the complexity and diversity of knowledge and tasks so that a wider array of learning and problem solving strategies need to be drawn on. By Sociology, Prestine and LeGrand (1991, p. 70) mean the social aspect of the learning environment. Expert practice is always developed 'in the context of its application to realistic problems within the culture of actual practice'. The notion of expertise in this context is characterized as 'the flexible reconstruction of prior knowledge' (Prestine 1993, p. 199) which, in turn, includes the four conceptions of expertise outlined by Kennedy (1987): expertise as technical skill, as the application of theory or general principles; as critical analysis, and as deliberate action, with particular emphasis on the last two.

The transfer problem, according to Prestine (1993, pp. 199–201), is expected to be overcome 'through externalizing expert cognitive processes' which would help integrate the new knowledge more meaningfully and thus make it more readily accessible in future and different problem-solving situations. These expert processes are internal but are said to be able to be documented by means such as protocol analysis and observation, and show themselves in modeling, coaching and scaffolding (Collins *et al.* 1989, p. 458). Prestine admits that this seems to be much easier to accomplish in well-structured rather than ill-structured domains such as education. Expertise in the latter will necessarily be a 'function of the complexity of the network of relationships among existing knowledge structures ... and the ability to control the process of flexibly reconstructing and reconstituting the relationships between the schema representations' (Prestine 1993, p. 201).

In the final sections we consider some of the critical issues raised by the various models of administrator preparation programs, with special emphasis on the last in terms of its assumptions of transfer, situatedness, human cognition, and the alleged distinction between *isolated and passive* knowledge on the one hand and *constructed active* knowledge on the other.

## The Problem of Transfer

Traditionally defined as 'grounded in individual accumulations of knowledge' (St. Julien 1997, p. 261), the classical view of the problem of transfer is phrased in terms of a learner's inability to use formal knowledge in another context when it would be appropriate to do so. 'In a sense', Resnick (1989, p. 8), observes, 'transfer is the holy grail of educators — something we are ever in search of, that hope pretends lies just beyond the next experiment or reform program.'

Following Greeno, Moore and Smith's (1993, pp. 160–161) classification, there are four theories of transfer, two of which have been highly influential in education and, by implication, in educational administration. These theories are (1) the classical-empiricist represented by the work of Thorndike and Woodworth; (2) rationalist theories of transfer as developed by Piaget and the Gestalt psychologists such as Duncker and Wertheimer; (3) the sociohistorical framework which refers to the work of situated action theorists such as Lave (e.g. 1997). The fourth framework is the ecological theory of transfer represented by writers such as Shaw *et al*.

The most influential in education, the empiricist and the rationalist theories of transfer, 'share the crucial assumption that transfer depends on the cognitive structure that the learner has acquired in initial learning and can apply in the transfer situation. They differ in their general approach to how these representations are acquired in initial learning' (Greeno *et al*. 1993, p. 161). The empiricist is interested in finding overlap of elements of components between two situations while the rationalist is concerned to find 'shared structure that a learner could carry over from initial learning to the transfer situation.' Chomsky's theory of language structures believed to be innate is an example of the rationalist account. The theory of transfer implicit in the sociohistoric perspective is based on the agent's 'having learned to participate in an activity in a socially constructed domain of situations that includes the situation where transfer can occur. Transfer depends on structure in the situation that is primarily socially defined, and that has been included in the person's previous social experience' (Greeno *et al*. 1993, p. 161). The ecological theory, while it resembles that of Lave and other social cognitivists, stresses mainly the structures in the physical environment.

Greeno, Moore and Smith (1993), as important representatives of the situated learning perspective, present an explicit account of transfer which combines features of both the sociohistorical and the ecological views. It is thus instructive to take a closer look at their theory since it helps to clarify further how transfer is expected to happen in the situated perspective and in the reform program suggested by Prestine especially.

Transfer is ultimately to be found 'in the nature of the situations, in the way that the person learns to interact in one situation, and in the kind of interaction in the second situation that would make the activity there successful' (Greeno *et al.* 1993, p. 100). Of particular significance are their much cited conceptions of *affordances* and *invariants*. Any activity in which a person is involved is shaped by properties of things and materials in the situation as well as by the characteristics of the person. Support for particular activities, as offered by the joint interaction of those things and materials, are what the writers call *affordances*. The characteristics of agents who are able to engage in these activities are their *abilities*. The relationship between *affordances* and *abilities* is one of reciprocal interaction. Greeno *et al.* (1993) claim that transfer from situation *s1* to situation *s2*

> involves a transformation of the situation and an invariant interaction of the agent within the situation … Transfer can occur if the structure of the activity is invariant across the transformation from *s1* to *s2* with respect to important features that make it successful, or if a needed transformation of the activity can be accomplished. … The activities that people learn are constructed and situated socially. To a great extent, the affordances that enable our activities are properties of artifacts that have been designed so those activities can be supported. The functions of these properties as affordances are shaped by social practices. People can learn these practices, including the utilities of affordances, mainly by participating in them along with other people. (Greeno *et al.* 1993, p. 102)

In aligning their views more with the ecological perspective they de-emphasize the discussion of cognition in its social and cultural contexts but stress that 'the understanding of cognition that we need' must take account of them.

One of the most important contributions Greeno *et al.*'s perspective on transfer makes is its emphasis and detailed examination of the nature of the situation and the agent's interaction within it. In this respect it departs importantly from the traditional transmission model which excludes any concerns for context, situation, or interactional features. Ironically though, while the traditional model believed that all relevant cognitive detail was located in a person's head, which in the case of Simon for example, consists of symbols and their manipulation, in the situated learning view, most if not all cognitive detail seems to be external to the agent, located in the situations and their interactions with things and materials (see Chapter 3). While it might be said that situated learning has contributed to help explain why knowledge does *not* transfer, it raises the problem of what exactly makes *positive* transfer possible (St. Julien 1997, pp. 264–265)?

Bereiter (1997, p. 286), too, notices this difficulty and states bluntly that situated cognition's main problem is 'precisely its situatedness' (see Kennedy's 1987, pp. 154, criticism of apprenticeship which makes a similar point). As we get more and more attuned to the constraints and affordances of a particular situation, he argues, we become less and less able to generalize what we have learnt to other situations, having moved from novice to expert in *s1* which is the general thrust of situated learning.

Being expert in *s1*, however, does not transfer to being expert in *s2* since the same progression from novice to expert has to be gone through in *s2* as well, as proposed by the assumption of developing increasing competence through peripheral or other activity in whatever situations agents find themselves. This, however, simply cannot be the case since intelligent behavior does indeed transfer (Bereiter 1997, p. 287). The example of space travel serves here in so far as it exemplifies 'our most colossal example of transfer of learning'. Bereiter (1997, p. 287) goes on, 'No amount of situated cognition or legitimate peripheral participation would get people to the moon and back. It took something more to produce that kind of transfer, and we must try to pin down what that is.'

Addressing himself specifically to the notion of transfer offered by Greeno and colleagues, Bereiter (1997, p. 288) argues that their account offers a valuable conceptual advance in that 'constraints and affordances are not characteristics of either the environment or of the person, considered separately, but of the relationship between person and environment. Thus, transfer is a matter of the same kind of relationship coming into play in different situations'. However, the problems it raises are quite significant.

For a transfer case to be noteworthy, the affordances and constraints of two situations must look different on the surface. Bereiter cites the well-known X-ray and crusader attack example of transfer which could only be accomplished once it was pointed out to the subjects that the first solution to the problem also provides the solution to the second problem. The connection agents could make between the two, once cued, was possible because they found the relevant relationship between them. And this is an *abstract* matter, 'a relationship based on formal, structural or logical correspondences' which can only be discovered if one creates 'symbolic representations of situations and carr[ies] out operations on those symbols.' (Bereiter 1997, p. 288). We return to this point shortly.

To the extent that administrator preparation programs such as the *cognitive apprenticeship* model discussed above assumes a mode of transfer as is developed by situated learning theorists (although only the conception of Greeno and colleagues is discussed here), the same problems can be raised. Despite the positive developments regarding the contextual nature of administrative practice, the issue of *positive* transfer, that is, learning, remains. Although it is true to say that abstract knowledge does figure in *cognitive apprenticeship*, as it does to some extent in the other reform programs mentioned, it is not clear how such knowledge is acquired, and how teachers, who are expected to draw out general principles, are able to do so since they themselves have presumably become experts the way one does as explained by situated learning, that is, through hands-on experience.

## Expert Practice

Since our discussion in this chapter is on how to understand expert practice and what makes it happen, it is useful to note what is known about expertise, what it consists in and how it is displayed. There are considerable difficulties in this field of study, beginning with how to know when expertise is in evidence since it only shows itself in a context of use. This was already indicated by Prestine when discussing the externalization of mental processes which are said to be indicative of the thinking of the expert in a particular situation.

Frensch and Sternberg's definition of expertise (as cited in Ohde and Murphy 1993, p. 81), as 'the ability, acquired by practice, to perform qualitatively well in a particular task domain', clearly shows the difficulties in determining when performance can be rated as expert performance, especially in social and ill-defined contexts. Since expertise only shows itself in a context of use, the question of expert knowledge is one of *how* prior knowledge comes to be used in practice in the way that it is. This issue is of considerable complexity. Ohde and Murphy (1993) present some of the major interrelated and apparently consistent findings of otherwise divergent studies. When comparing the performance of an expert to that of a novice across many fields, the following findings appear to be shared features (Ohde and Murphy 1993, pp. 75–76):

(1) An expert within a specific domain will have amassed a large yet well-organized knowledge base …
(2) This extensive body of knowledge allows experts to classify problems according to principles, laws, or major rules rather than surface features found within the problem …
(3) The knowledge base is highly organized, allowing experts to quickly and accurately identify patterns and configurations. This ability reduces cognitive load and permits the expert to attend to other variables within the problem …
(4) The problem-solving strategies of experts are proceduralized. Experts can automatically invoke these skills while novices often struggle with the problem-solving process …
(5) The acquisition of this complex knowledge base takes a long time. Expertise within a domain is linked to years of practice, experience, or study…

Expert practice, then, takes a long time to develop (a point also strongly supported by Simon 1980), is characterized by an apparent lack of conscious effort, and seems fluid and intuitive. Experts possess a lot of knowledge about their respective domains of activity including well-honed appropriate skills. Following Dreyfus and Dreyfus' model (1986; Dreyfus 1987), the development of expertise from novice to expert can be thought of as happening along a continuum in terms of degrees and stages which does not necessarily issue in every novice becoming an expert.

It is now time to attempt to unravel some of the arguments and assumptions of the situated learning perspective and their implications for reform models of administrator preparation programs which we think lead to unnecessary confusion. The issues most important to discuss are (1) the active/passive distinction of knowledge acquisition; (2) the question of knowledge representation; (3) the conception of 'situatedness', and finally, the so-called problem of transfer.

## The Active/Passive Distinction of Knowledge Acquisition

The distinction between passive and active knowledge which maps on to the distinction between transmitted, taught, abstract knowledge versus constructed, active, and practical knowledge, in our view, is mistaken. Depicting the learner as either an active constructor or a passive recipient of knowledge is inaccurate since knowledge *acquisition* is always 'active' in that human beings 'construct' their knowledge of the world in which they grow up. This is a point we have made repeatedley (Evers and Lakomski 1991, 1996), but it is one which bears re-emphasizing. Human beings cannot but construct their knowledge since they are not born with any but tend to end up knowing quite a lot. This of course includes knowledge of words and language which the child acquires gradually, and only partly by being told. There is no in principle difference between the child's acquiring the words 'cat', 'mum' and 'hungry' and the adult administrator's learning words such as 'budget', 'downsizing', and 'case study'. Both 'construct' the meaning of these terms within the context of their already existing theory of the world, or fail to do so if the new knowledge is too remote from what they currently know.

Without wanting to engage in a debate about what is called *constructivism* in its various forms and guises — the reader is referred to the excellent analyses by Phillips (1995, 1997) — the confusion we detect in the preparation literature discussed is between pedagogy (methodological) and nature of learning and cognition (epistemological). Stated differently, what is primarily at issue is not whether or not knowledge ought to be taught in the lecturing mode, or 'discovered' in practical activity, concerns amply discussed in Deweyan progressive education, *but how knowledge is represented.* This is not to say that the pedagogical issues are unimportant, but it is to stress that what causes the confusion, and hence difficulties, in explaining why transfer or learning does or does not happen, is the absence of an account of human knowledge representation *in both symbolic and sub-symbolic forms.*

A first step to the solution of creating better administrative practice is the realization that knowledge of practice is not representable in symbolic form. The active/passive distinction, in the form in which it does most damage, that is, as a mistaken assumption about the nature of human learning, just dissolves. Once this is realized, the issue of professional competence and expertise can be addressed more constructively. Specifically, the issue now becomes one of attempting to clarify the function of symbolic representation, such as is exemplified by lectures and in textbooks, in light of our knowledge of human knowers as pattern recognition and completion specialists, whose brains are more analogous to vast and multiply interconnected nets than they are to linear computers (as we argued in Chapter 3). The neural net model of knowledge acquisition and processing which drives our naturalism accounts for knowledge in terms of the neuronal weights, or connection strengths, between the nodes which constitute patterns of activations. Learning happens when weights change, either because of internal reconfigurations or the effects of external environmental inputs, or a combination of both. Amazingly, the human brain manages to complete an incomplete pattern quite satisfactorily as in seeing someone's nose and realizing it belongs to your sister. A pattern is activated when enough of its nodes are present, but not all of them have to be present. This means that different nodes in different circumstances and at different times can cause the representation of the same pattern of activation. We can think of a thing, a person, an incident, or a concept as being the same but do so by having activated a sufficient number of nodes quite different from another activation occasion, provided their connection weights are strong enough to activate the pattern.

This human capacity makes possible sewing a garment expertly as it does solving a mathematical equation. It is important to stress here that neither the practical skill of expert sewing nor the symbolic task of solving equations are *stored* in the brain as some kind of data structures. When the tailor or the apprentice are not sewing, there is no 'pattern of trouser sewing', no template, left in the brain, and this equally applies to the mathematical problem-solver. Particularly with regard to the latter, it is not the case that symbols such as words or numbers are stored *as* words or numbers. In a connectionist system such as the brain there are only neuronal connections which may get activated into patterns when the appropriate weights are present and are strong enough. A symbol *is* an activation pattern, and when the activation pattern is not active then the symbol and its neuronal connections dis-assemble, with nothing left over.

It is because of this fundamental feature of the connectionist system that is our brain that the distinction between active and passive knowledge acquisition evaporates. So when we say that practice makes perfect, we can translate that into its proper naturalistic frame. The competence or excellence we exhibit when sewing a garment, cooking a meal, solving mathematical problems, or leading a school, is attributable to the strengthening of the connection weights between the nodes which cause the pattern to come into being. This also applies to moving from novice to expert status, as nicely illustrated with an example Simon provides. Unlike Simon (1980, p. 84), however, who argues that our long-term memory can be thought of as 'a large indexed and cross-referenced encyclopedia in which all articles are arranged irregularly…', becoming an expert X-ray diagnostician is not a matter of storing 'chunks' of knowledge in memory, but of seeing many X-rays over a very long period of time. This means that the requisite activation patterns, by repetition, become strengthened in terms of the connection weights between their nodes. We might say that the diagnostician's brain becomes finely tuned to all the potential disease patterns which lungs can exhibit. Such expert knowledge then shows itself as expert knowledge commonly does: intuitively, instantly, and without apparent reflection or deliberation. If this is so, how do we account for language as a symbol system in our naturalistic account of practice?

## The Question of Knowledge Representation

In order to answer this question we have to draw a distinction between language at the individual and language at the cultural level. This move will also involve us in discussing some recent developments in connectionism which throw further light on the function of language. In an abbreviated form we can describe the function of language for both individual and social contexts as a kind of *compression algorithm of experience*. The discussion which follows takes us simultaneously into the area of situatedness and how our naturalism explains it.

The significance and function of symbol systems such as the linguistic one is an issue still very much to be explored in connectionism. We pointed out earlier (in Chapter 3) that the significance of formalized knowledge in linguistic form goes beyond language as the means of intra- and interpersonal *communication*. It also serves the broader function of making human cognition *collective* (Churchland 1995). Considered from an individual (as separate from the cultural) point of view, formalized knowledge in linguistic structures, in being publicly available, makes it possible for agents to check on the validity of statements and claims. Since it is formalized it is universally applicable and this has the advantage that agents can avail themselves of its contents without needing to have had the actual experience which it represents. This point is quite important in so far as agents are finite biological bodies, located in time and space, with access only to their immediate environments and experiences. Formalized knowledge represented in linguistic form opens up the experiences of others, *past and present*, and allows us to speculate about the future as well, as in making plans, writing mission statements, shaping policies, etc. This would seem a rather important point to make in relation to the use and utility of textbooks and formal lectures in any preparation program for administrator or teacher, since our own personal experience is always limited by nature. When we shift our attention to knowledge formalization at the cultural level, however, things look different. And this is the level with which we need to be concerned since individuals are also always social beings, shaped by the culture into which they are born and which they shape in turn, including its institutions, artifacts, materials and objects. Happily, this is the direction contemporary connectionism is taking. (For a first naturalistic exploration of culture from within the discipline of anthropology see Strauss and Quinn 1997).

These new developments derive from what might be called revisionist cognitive history — the alternative history of cognitive science as sketched by Hutchins (1996) for example — which filled in the gulf between individual cognition, confined in one skull, and the outside world created by the assumptions and long-lasting dominance of traditional Artificial Intelligence as represented in the physical-symbol system hypothesis. We have discussed these developments in some detail earlier, so let us summarize its most important results for the purposes of the present discussion.

Putting brain, body and world together again, the apt sub-title of Clark's (1997) book, discusses the attempt to explain how cognition is *distributed* in the world. The basic idea is that since humans have limited computational power, it is necessary for us to contract out complex cognitive tasks to lighten the brain's load. This contracting out is just nature's clever way of conserving energy and avoiding cognitive overload. It is also clever because it has allowed us to develop very sophisticated machines, processes, and organizations — culture — which would not have come into being without the collective outsourcing and harnessing of computational horse power. But the structures we created are not inert. Because of the way they have been composed, in banks, universities, and hospitals, for example, they exert their own dynamics in return. We have devised schools, for example, which consist of classes and timetables, and teachers, students, parents and administrators comply with these structural entities in terms of organizing their lives around them. This is a mundane example, and much of our everyday lives' organization looks mundane until we begin to question why it works the way it does and try to find out what cognitive efforts are involved in maintaining it. Cultural connectionism provides a first attempt to consider the external material structures of everyday lives as the embodied and embedded extensions of our minds. (An excellent example is Hutchins' 1995 discussion of what happens in a cockpit).

Language, to return to our earlier theme, is just such a means of cognitive outsourcing. Since we cannot memorize everything we experience, off loading memory in written texts is an important function of language; also, since there are things we cannot experience, written documents allow us to have access to those experiences as others had them so that experience and the results of the cognition of others are not lost; we devise labels as cognitive shorthand by means of which we maneuver around our environments more or less successfully, take traffic symbols and road signs as obvious examples; importantly, too, language helps us coordinate activity with other agents; and when we hold inner monologues we actually direct our own attention internally in certain directions. The important result of these considerations is that the externalization of individual thought through language explains why it is possible to have culture at all because it transcends any individual's experience and joins it with those of others. The cognitive resources thus created are immense, grow exponentially through a kind of cognitive bootstrapping, and are in principle unlimited.

How do these reflections relate to the preparation of prospective administrators? The obvious conclusion to state is that textbooks and other linguistic forms of instruction such as lectures are serving an important function in that they make available cultural experiences the novice has not had. It is 'stored' knowledge about specific domains including theories as particularly formalized knowledge structures. The abstractness or generality of knowledge decried by Murphy, Hallinger, Bridges, Prestine and other writers in the situated learning perspective is thus abstract in the sense of being a compression of others' cognitions which in their formalized linguistic structures survive the actual knowers who thought them. Hence, abstractness or generality of knowledge has nothing to do with the location of its origin. This is a confusion which ends up costing reform programs valuable cognitive resources to the extent they de-emphasize their use. It needs to be remembered that symbolic representations are also patterns based on neuronal connections, it's just that they are, unlike human practice, that 'face of reason' which we *know* how to represent.

In practice, reformers do not actually advocate eliminating textbooks or 'taught' knowledge altogether, but rather suggest that prudent use should be made of them in the context of simulations, case studies, or critical incidents. Our account of naturalized practice tells them why it is indeed a sensible thing to do in light of how we acquire and process information in social contexts. It is an account such as this which is missing in the situated cognition perspective.

When we consider the almost exclusive emphasis on *situated* learning as argued for in the cognitive apprenticeship model, we can now say that practical knowledge and the development of expertise is not superior because it is locally developed knowledge. This claim is too strong because it is *not the local character of knowledge* which is at issue. What is at issue is primarily the capacity to learn, and how such learning shows within the environments in which the learner is causally enmeshed and which exert their own pull. The distinction is not a trivial one since although there are differences in emphases in the situated learning field, the role of human cognition in interaction with its environment has not been developed, as we saw by way of discussing the conception of transfer by Greeno and colleagues who stressed ecological factors over others but recognized the importance of social cognition. (For a recent extended discussion of situated cognition see Clancey 1997).

## Transfer, Situatedness, and How It's All Connected

Finally, what of transfer? In light of the previous discussion, the so-called problem of transfer as traditionally defined can be suitably re-framed. The issue is no longer one of appropriate transfer of abstract knowledge learnt in context $A$ to successful application in context $B$. In fact we can leave out the term 'transfer' altogether and ask: were the appropriate activation patterns activated in the new context or not? If they were not, then the agent might not have learnt a specific concept in the first place so that there were no patterns *to* activate, or the existing connection strengths were not sufficient to activate or complete the pattern. This is a rather global answer, but we believe that it indicates the right direction to follow if we want to understand why students or prospective administrators might not learn, or demonstrate learning in specific contexts. Furthermore, the empirical results obtained in novice/expert studies make *cognitive* sense in that the development of expertise implies (a) a lot of knowledge, and (b) a long time frame. It also follows from our discussion that competence or expertise is situation-bound. Because of the intricate interplay of individual cognition and external structure, an interplay which differs more or less subtly even between similar situations, some may be competent administrators in one context and yet fail in another because of the differences in the new environment which may activate or fail to activate the requisite patterns. A 'transformational' leader may be successful in one school but fail to bring about the kind of change required in another, and this is just what one would expect.

Returning to March's astute observations, we are now in a position to state that any preparation program needs to be mindful of the fact that administrator preparation is a social act, not in the commonsense understanding of the term, but because individual brain states are themselves socially or collectively shaped. The structures, artifacts, and objects which surround the novice administrator are as important as are the books, articles and other documents which are part of the program, however their relative distribution is determined in practice. This is another way of saying that practice-based administrator preparation, because it is situation-specific in the sense of having available the very structures, etc. which make up the normal environment of administrators, is to be preferred, including use of those texts which represent the best recent knowledge pertaining to the business of administration broadly conceived. Our naturalistic approach thus makes it possible to unite administrative theory and administrative practice. In so doing, it makes available for the first time a unified conception of administrator preparation which is able explicitly to draw on *all* our socially and physically embedded knowledge. Such an expanded conception, we believe, provides the most fertile framework yet for the planning of preparation programs that fully take into account the seamless web of administrative experience.

# References

Anderson J.R., Reder L.M. and Simon H.A. (1996). Situated learning and education, *Educational Researcher*, **25** (4), 5–11.

Argyris C. and Schön D.A. (1996). *Organizational Learning II*. (Reading, MA: Addison-Wesley).

Bereiter C. (1997). Situated cognition and how to overcome it, in D. Kirshner and J. A. Whitson (eds.) *Situated Cognition*. (Hillsdale, NJ: Lawrence Erlbaum Associates).

Bridges E.M. and Hallinger P. (1995). *Implementing Problem Based Learning in Leadership Development*. (University of Oregon: ERIC Clearinghouse on Educational Management).

Churchland P. (1995). *The Engine of Reason, The Seat of the Soul*. (Cambridge, MA: The MIT Press).

Clancey W.J. (1997). *Situated Cognition*. (Cambridge: Cambridge University Press).

Clark A. (1997). *Being There: Putting Brain, Body, and World Together Again*. (Cambridge, MA: The MIT Press).

Collins A., Brown J.S., and Newman S.E. (1989). Cognitive apprenticeship: teaching the crafts of reading, writing, and mathematics, in L. B. Resnick (ed.) *Knowing, Learning, and Instruction*. (Hillsdale, NJ: Lawrence Erlbaum Associates).

Dreyfus H.L. and Dreyfus S.E. (1986). *Mind over Machine*. (New York: The Free Press).

Dreyfus H.L. (1987). Misrepresenting human intelligence, in R. Born (ed.) *Artificial Intelligence*. (London and Sydney: Croom Helm).

Ernst G.W. and Newell A. (1969). *GPS: A Case Study in Generality and Problem Solving*. (New York: Academic Press).

Evers C.W. and Lakomski G. (1991). *Knowing Educational Administration*. (Oxford: Elsevier).

Evers C.W. and Lakomski G. (1996). *Exploring Educational Administration*. (Oxford: Elsevier).

Frensch P.A. and Sternberg R.J. (1989). Expertise and intelligent thinking: when is it worse to know better?, in R.J. Sternberg (ed.) *Advances in the Psychology of Human Intelligence*: Volume 5. (Hillsdale, NJ: Lawrence Erlbaum Associates).

Greenfield T.B. and Ribbins P. (eds.) (1993). *Greenfield on Educational Administration*. (London and New York: Routledge).

Greeno J.G., Moore J.L., and Smith D.R. (1993). Transfer of situated learning, in D. K. Detterman and R.J. Sternberg (eds.) *Transfer on Trial: Intelligence, Cognition, and Instruction*. (Norwood, NJ: Ablex Publ.).

Hutchins E. (1995). How a cockpit remembers its speeds. *Cognitive Science*, **19**, pp. 265–288.

Hutchins E. (1996). *Cognition in the Wild*. (Cambridge, MA: The MIT Press).

Kennedy M.M. (1987). Inexact sciences: professional education and the development of expertise, in E. Z. Rothkopf (ed.) *Review of Research in Education*, **14**. (Washington, DC: American Educational Research Association).

Lave J. (1997). The culture of acquisition and the practice of understanding, in D. Kirshner and J.A. Whitson (eds.) *Situated Cognition*. (Hillsdale, NJ: Lawrence Erlbaum Associates).

Leithwood K., Begley P.T., and Cousins J.B. (1994). *Developing Expert Leadership for Future Schools*. (London and Washington, DC: The Falmer Press).

March J. G. (1974). Analytical skills and the university training of educational administrators. *The Journal of Educational Administration*, **XII** (1), pp. 17–44.

Mintzberg H. (1973). *The Nature of Managerial Work*. (New York: Harper and Row).

Mintzberg H., Raisinghani D., and Théorét A. (1976). The structure of 'unstructured' decision processes. *Administrative Science Quarterly*, **21**, pp. 246–275.

Murphy J. (1992). *The Landscape of Leadership Preparation*. (Newbury Park, CA: Corwin Press).

Nisbett R.E. and Ross L. (1980). *Human Inference: Strategies and Shortcomings of Social Judgment*. (Englewod Cliffs, NJ: Prentice-Hall).

Ohde K.L. and Murphy J. (1993). The development of expertise: implications for school administrators, in P. Hallinger, K. Leithwood, and J. Murphy (eds.) *Cognitive Perspectives on Educational Leadership*. (New York and London: Teachers College Press).

Phillips D.C. (1995). The good, the bad, and the ugly: the many faces of contructivism, *Educational Researcher*, 24(7), pp. 5–13.

Phillips D.C. (1997). Coming to grips with radical social constructivism. *Science and Education*, **6**, pp. 85–104.

Prestine N.A. (1993). Apprenticeship in problem-solving: extending the cognitive apprenticeship model, in P. Hallinger, K. Leithwood, and J. Murphy (eds.) *Cognitive Perspectives on Educational Leadership*. (New York and London: Teachers College Press).

Prestine N.A. and LeGrand B.F. (1991). Cognitive learning theory and the preparation of educational administrators: implications for practice and policy. *Educational Administration Quarterly*, **27** (1), pp. 61–89.

Resnick L.B. (1989). Introduction, in L.B. Resnick (ed.) *Knowing, Learning, and Instruction*. (Hillsdale, NJ: Lawrence Erlbaum Associates).

Robinson V.M.J. (1993). *Problem-based Methodology: Research for the Improvement of Practice*. (Oxford: Pergamon Press).

Robinson V.M.J. (1998). Methodology and the research-practice gap, *Educational Researcher*, 27(10), pp. 17–26.

Simon H.A. (1945). *Administrative Behavior*. (New York: The Free Press, 3rd Edition).

Simon H.A. (1977). *The New Science of Management Decision*. (New York: Harper and Row).

Simon H.A. (1980). Problem solving and education, in D.T. Tuma and F. Reif (eds.) *Problem Solving and Education: Issues in Teaching and Research*. (Hillsdale, NJ: Lawrence Erlbaum Associates).

Sternberg R.J. and Frensch P.A. (1993). Mechanisms of transfer, in D.K. Detterman and R.J. Sternberg (eds.) *Transfer on Trial: Intelligence, Cognition, and Instruction*. (Norwood, NJ: Ablex Publications.).

St. Julien J. (1997). Explaining learning: the research trajectory of situated cognition and the implications of connectionism, in D. Kirshner and J.A. Whitson (eds.) *Situated Cognition*. (Hillsdale, NJ: Lawrence Erlbaum Associates).

Strauss C. and Quinn N. (1997). *A Cognitive Theory of Cultural Meaning*. (Cambridge: Cambridge University Press).

Tversky A. and Kahneman D. (1983). Extensional versus intuitive reasoning: the conjunction fallacy in probability judgment, *Psychological Review*, **90**, pp. 293–315.

Tversky A. and Kahneman D. (1986). Rational choice and the framing of decisions, *Journal of Business*, **59**, pp. 251–278.

Wagner R.K. (1993). Practical problem-solving, in P. Hallinger, K. Leithwood, and J. Murphy (eds.) *Cognitive Perspectives on Educational Leadership*. (New York and London: Teachers College Press).

# PART III

# Researching Practice

# 9

# Coherent Educational Research

While in previous chapters we have drawn on a variety of research findings, especially about the new cognitive science, to develop a perspective on the practice of educational administration, we have said little about how these findings bear on the practice of research itself. In this last part of the book we attend to this matter. The present chapter provides a general orientation to the main features of our coherentist epistemology as they apply to the nature of educational research. In view of the comparative and survey genre of the discussion, which deals with mainly symbolic representations of knowledge, we formulate our coherentism at this higher level perspective. A move to the lower level naturalism that our position endorses and coheres with, occurs in Chapter 10, where we apply our general framework to issues bearing on researching practice. This chapter begins therefore by offering an overview of key topics in educational research methodology as seen from our naturalistic and coherentist approach to knowledge and its justification.

The development of research methodologies is a way of formulating and making explicit sound procedures for inquiry — procedures that determine the nature of acceptable evidence, the kinds of inferences that may be drawn and, as exemplified in the praxis tradition, the kind of action that is appropriate. Construed this way, methodologies are influenced primarily by epistemological assumptions, particularly by assumptions about whether and how knowledge is justified. In prescribing canons concerning the nature and limits of justification, epistemologies exercise a normative function, directly over methodology, and indirectly over the structure and content of substantive theories purportedly sustained by research.

In the discussion that follows, three main stages in educational research methodology that have been associated with major epistemological positions, will be explored. We argue that all three stages reflect characteristic limitations associated with foundationalist assumptions in knowledge justification. As an alternative, a coherentist model of knowledge justification will be outlined, and its incorporation into educational research methodology recommended. The coherentist model to be outlined is a species of naturalistic epistemology, that is, one informed by natural science accounts of how humans acquire knowledge and process information. One consequence of this approach is a blurring of the so-called quantitative/qualitative distinction. Another is a blurring of the alleged distinction between natural science and social science. The main result will be an approach to inquiry that is somewhat more inclusive in what it counts as knowledge, but also more discriminating in discerning what should be left out.

## Stages of Educational Research Methodology

It is convenient to categorize mainstream educational research methodology into three broad stages.

Although these stages are partly chronological in this very general characterization, they grew out of successive attempts to theorize inquiry, and all flourish today. A view of methodology as comprising a range of paradigms is still the dominant perspective in the field. A more detailed account of each, together with an account of their key interrelationships follows.

### Traditional Science of Educational Research

Traditional quantitative scientific approaches to educational research, once seen as the only proper way to conduct inquiry, but now mostly regarded as just another paradigm, may be distinguished by three structural features that reflect their background logical empiricist epistemological assumptions.

On the question of *justification*, a partition on knowledge claims is posited which demarcates an epistemically privileged subset from all the rest. This subset, usually observation reports, plays a foundational role in justification, with justification proceeding by way of testability, which has two components. Thus, non-privileged claims within a theory are thought to imply privileged ones. If these obtain, the theory is confirmed, and is said to be supported by evidence. If contrary observations obtain, the theory is said to be disconfirmed and in need of revision. A theory is more

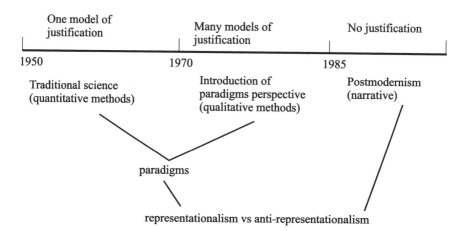

**Figure 9.1** Three stages of educational research methodology.

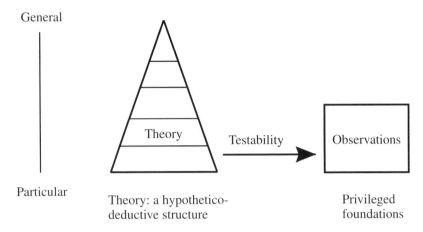

Figure 9.2 Traditional science of educational research.

justified than its rivals if it has amassed more confirmations and fewer or no disconfirmations (For a standard account, see McMillan and Schumacher 1993, pp. 78–90).

Because of the demands of testability, theories need to exhibit a *hypothetico-deductive structure* (Kerlinger 1986, pp. 15–25). Since implication for testability runs from claims that are more general to claims that are less general, this structure will be a hierarchy of statements, with unrestricted law-like generalizations at the top, descending to singular claims at the bottom. Educational theorizing, because it is so context specific, is located in the intermediate layers of generality, such theories sometimes being called 'theories of the middle range'. They are justified in the usual way, by the deduction of claims that belong to the privileged subset for subsequent testing, but can also be justified if they follow from networks of warranted claims located further up the theoretical hierarchy. For this reason, some branches of educational studies (e.g. Educational Administration, Educational Psychology) appear to be influenced more by a generic parent discipline (e.g. Management, Psychology) than by Education.

The third structural feature of traditional science approaches to research is quite technical, residing in the background as an assumption of more rigorous formulations of logical empiricism. It concerns the language in which theories are to be formulated. Owing to the influence of Vienna Circle logical positivists of the 1920s and 1930s, scientific theories were thought to be best formulated in an austere logical language which had precise syntactical rules for forming valid symbol combinations and a precise semantics which displayed in unambiguous fashion the meaning of each sentence. The logical notation for which this had easily been achieved by the 1950s was the predicate calculus, developed at the turn of the century by Russell and Whitehead.

In understanding the implications of logical empiricist driven methodologies for the social sciences it is important to know that the predicate calculus, if used as an ideal language for theory construction, is *referentially transparent,* or *extensional* (Linsky 1971). What this means can be illustrated by the following example:

Consider the argument

    (1)  7 is less than 9
    (2)  9 = the number of planets
    (3)  7 is less than the number of planets.

Provided premise (2) is true, we can correctly deduce the truth of (3) from the truth of (1) and (2). Now consider a corresponding argument.

    (4)  Jones knows that (7 is less than 9)
    (5)  9 = the number of planets
    (6)  Jones knows that (7 is less than the number of planets).

This argument for (6) is invalid even given the truth of (4) and (5).

There is a crucial difference between the two arguments. The first is valid regardless of how the number nine is referred to. Provided the different expressions '9', or 'number of planets', or 'number of my second cousins' or 'number of coins in my pocket' all refer to the same object, in this case the number nine, any of these expressions can be substituted into the argument without affecting its validity. In the predicate calculus, all expressions that refer to the same thing can be exchanged without affecting the truth of the sentence which contains them. That is, this ideal language is referentially transparent. However, the second argument contains sentences that are referentially opaque. What Jones can be said to know depends importantly on how it is referred to. And this is the case not just for Jones's knowledge, but also for Jones's beliefs, desires, hopes, wants, imaginings, and all the rest of what are called the propositional attitudes. To restrict the damage referential opacity does to the job of constructing a human science, some way must be found for limiting the range of semantically valid descriptions of human conduct. The usual way has been to insist on giving operational definitions for any propositional attitudes that turn up in research. The effect is to restrict referring expressions to unambiguously specified, observable, measurable behaviors.

Whatever the merits of logical empiricist based accounts of research in natural science, they constitute an exceptionally severe restriction on the data of social science. Empiricist theory of evidence rules out familiar appeals to human subjectivity — including such matters as the individual construction of meaning and the interpreted significance of events — and also the relevance of ethics as an integral constraint on inquiry. The hypothetico-deductive model of theories, while useful for dealing with deductive relations among context invariant generalizations of the sort that can be found in theoretical physics, becomes rather brittle and implausible as a model for the highly context dependent social phenomena of education where there may be no true, non-trivial, law-like, generalizations. And finally, the background demand for an extensional ideal theoretical language rules out much of the vocabulary that is used to characterize human activity and conduct, filtering intentional, meaningful, action down to a set of intersubjectively measurable behaviors.

So restrictive are the methodological demands of traditional quantitative research approaches that they have seemed to many to be inapplicable beyond natural science, prompting belief in a fundamental methodological bifurcation between the natural and the social world. Indeed, this bifurcation has been virtually institutionalized by the widespread adoption of a paradigms view of research.

*Paradigms of research*

The abandonment of logical empiricism in philosophy was occasioned not so much by its inability to account for social science as its obvious deficiencies in accounting for theory choice among competing theories in natural science. As we noted in Chapter 1, work in the 1950s by philosophers such as Quine, Hanson, Feyerabend, and Kuhn, raised three major difficulties for traditional empiricist accounts of knowledge justification in science.

First, there is no such thing as so-called 'hard data'. All observation reports in any epistemically favored set of claims involve some interpretation, and are described in some theoretical vocabulary, even for the articulation of measurement procedures used in operational definitions. Experience itself is filtered by theory, and so too is the identification of epistemic privilege (Petrie 1972). Second, many different theories can imply the same finite set of observations, thus making problematical the question of which competing theory is actually being confirmed. Finally, where a conjunction of many hypotheses is required to deduce an observation report, it is not always clear precisely which hypothesis or group of hypotheses a disconfirming observation is falsifying. Empirical test situations are always complex.

The influential conclusion drawn by Thomas Kuhn (1970) was that empirical evidence is always insufficient for rational theory adjudication. Assuming that empirical evidence is all the evidence there is for theory choice, and knowing that choices have been made in the ongoing development of science, he concluded that other, non-rational factors, were involved. Indeed, for Kuhn, the whole notion of some principle of rational choice extending beyond the influence of the comprehensive theoretical perspectives he called paradigms, was suspect.

Translated into an account of research, this means that questions about the relations between theory and evidence, and what counts as justification, are taken to be internal to comprehensive methodological perspectives. That is, there is no external methodological framework to which appeal can be made beyond particular paradigms of research, to rule on the appropriateness of paradigmatic differences. There are just different and distinct paradigms of research. With multiple models of justification in the offing, the matter of theory structure also fragments along paradigmatic fault lines, since theory of justification is a major determinant of the structure and content of substantive educational theories. Hypothetico-deductive structures need function as a constraint only on theories in natural science, while the demands of understanding, or empathy, or insight, or the quest for meaning, can impose their own distinct demands. Semantical constraints on the language of theory formulations vary with paradigms too. For example, the theoretical motivation in logical empiricism for an austere referential semantics assumes no access to the lush domain of meaning that is part and parcel of our everyday communicative practices. Yet even to mount a case for austerity assumes a certain amount of the antecedent lushness that is unquestioned or perhaps celebrated in other paradigms.

More than any other stage of research, the paradigms view has been responsible for a vast opening up of knowledge in educational studies. Much of the diversity that is now commonplace in the field owes its existence to the anti-empiricist assumptions that underwrite multi-perspectivism. Indeed, it is hard to imagine today's flourishing of varieties of subjectivism, interpretivism, critical theory, and cultural studies, to name just a few of the current options, without a corresponding attack on traditional science and its purported methods (Husén 1997).

Interestingly, while all this diversity might be thought dramatically to reduce the amount of screening performed by the paradigms research filter, the claim that the different paradigms are distinct, orthogonal, or better, totally independent, has a curiously opposite effect. For just as stage one methodology discounted the value of any research done outside traditional science methods, many of the paradigms return the compliment by discounting, or filtering out, natural science. The effect is odd. Consider, for example, what is required to interpret a measurable, intersubjectively observable sequence of behaviors in context, as a person teaching a class. Either appeals to interpretations, and interpretations of others' interpretations, the analysis of shared meanings, and a grasp of understandings are regarded as part of a methodologically complete and autonomous level of explanation or they are not. If yes, then it needs to be noted that all the physical accompaniments of teaching, such as gesturing, speaking, organizing, or writing, are accomplished by a physical body enmeshed in a causal field, presumably explicable by some other level of explanation that makes use of talk of nerve pathways, the firing of neurons, and the activity of arm muscles. The puzzle is that if the autonomy of paradigm-relative explanation is true, it is something of a miracle for the body to gesture, or move appropriate speaking muscles, at precisely the time that a person intends to teach some matter. The point generalizes, rendering the link between thought and action entirely mysterious. On the other hand, if paradigms do not function to filter out alternative perspectives and some coherent interlevel meshing is required to avoid the miraculous coincidence problem, then the distinctiveness of paradigms is compromised in a major way. (See Chapter 2 for more on the problem of meshing levels of explanation.) To date, these issues have not been squarely addressed in the methodology literature. (For a philosophical treatment, see Cussins 1992).

## Research as narrative

Epistemological attacks on logical empiricism that gave rise to the acceptance of paradigms in research raised serious doubts about the adequacy of appeals to empirical foundations to justify knowledge. The upshot has been a shift towards relativism and a diversity of ways of forming and defending theoretical representations of educational reality. Since the mid-1980s, however, a third stage of research has developed drawing rather more skeptical conclusions from the arguments against foundationalism.

Under the influence of postmodernism, three doctrines have gained prominence. The first is *anti-foundationalism*: there are no foundations to knowledge, and hence the justificationist problematic should be abandoned altogether. The second is *anti-essentialism*: concepts have no essential defining features, or things no corresponding essence. The third is *anti-representationalism*: theories, and the contents of our mind, do not mirror the world, or represent the way the world is. They fail to do so because there is no such thing as accurate representation (Evers and Lakomski 1996, pp. 262–270; Constas 1998).

For methodology framed within these doctrines, matters of truth and evidence drop out, along with the fiction/non-fiction distinction. Research ceases to be a quest for truth, or an attempt to build up a warranted representation of the world, but becomes rather an exercise in story-telling, in producing a narrative, and in giving voice to different viewpoints. Recorded data are more like texts whose meaning at first reflects the understandings the researcher has of the experiences of others, and then reflects the understandings of the reader. Outside of the justificationist problematic, the structure of theory is dictated by the requirements of narrative, with theory formulation characteristically being symbolic expressions of language (Connelly and Clandinin 1990; Clandinin and Connelly 1996).

This transition from representational to non-representational conceptions of knowledge and research has consequences for research as social practice. For example, within the praxis tradition, the notion of integrating normatively directed change with inquiry still presupposes theories that in some way represent the world, even if that representation reflects only one of many different possible perspectives. That is, the theorizing of change itself is in terms of a dynamical representation of researchers and their material contexts. But for theories developed in a praxis tradition detached from the idea of representation, the conceptualization of change is weakened, being limited mostly to theoretical claims about voice. However, since these theoretical claims in turn are not usually thought of as being subject to a representationalist normative ideal of truth, the virtues of voicing non-representational research are not easily formulated, least of all within the language of change consequences. Non-representational theories of voice, value, politics, and social change, are particularly problematical as frameworks for the analysis of research as (non-representational) narrative. There are numerous reasons for this, but a basic one is the inability of postmodernist research approaches in general to function in an importantly discriminating way about certain features of human experience.

If human behavior were totally random, no language would be possible, nor interpretation, nor understanding. For creatures inhabiting an environment in which there are limited resources, there are advantages in forming trajectories which lead to the solution of problems at better than chance. Driving a car, crossing a busy road safely, leaving a room through the door, making a cup of coffee, and performing a thousand other mundane physical and social acts is possible only against a background of sustained patterned, non-random, activity. The challenge to act on experience in a way that is more epistemically progressive than tossing a coin, is the challenge to solve what we might call the navigation problem. It is the task of devising non-arbitrary strategies for meeting needs, achieving goals, solving problems, and in general, getting around in the world while avoiding coming to grief sooner rather than later.

Part of the cognitive apparatus required to solve the navigation problem, to the extent that it is solved, is possession of a reliable map. Less metaphorically, navigating a successful trajectory is a matter of having a causally efficacious representational structure that contains epistemically useful information. Thus when a dog succeeds in uncovering the bone it buried, or when a bat utilizes acoustical evidence to fly between the bars of a cage, or when an infant negotiates its way around obstacles to obtain a favored toy, or when an adult plans to make an organization more effective, giving a complete explanation of successful outcomes of these actions will involve reference to the adequacy of the organism's capacity to represent salient features of its environment to itself. A methodology which does not permit one to formulate the notion of filtering out failures from successes in this modest, piecemeal, sense will have little of value to contribute to the sciences — natural, social, or human.

## Naturalistic Coherentism

Any proposal to address the weaknesses canvassed in the three mainstream stages of educational research methodology will need to meet a number of conditions. First, to perform useful work as a discriminating filter at all, a plausible distinction between justified and unjustified knowledge claims needs to be defended against the known difficulties with foundationalism. Second, some inclusive account of the structure of substantive theories needs to be given, especially one able to deal in a less brittle way with the graded, limited, and context dependent nature of social science generalizations. And finally, since much research in education concerns *practices,* some way of formulating practical knowledge needs to be found. It is claimed here that an approach to knowledge and its representation that incorporates elements of both naturalism and coherentism can provide an adequate framework for the development of more suitable research methodologies in educational studies (Evers 1991).

### The paradox of inquiry

One important test for any approach to research methodology is its capacity to deal with what is known as the 'paradox of inquiry'. Formulated by Plato in the *Meno*, it runs roughly as follows: either we know what it is that we are inquiring after, in which case inquiry is unnecessary, or we do not and so would not recognize it if we found it, making inquiry pointless; therefore inquiry is either unnecessary or pointless. Of the many possible responses to this paradox, it is useful to begin with one of the criticisms of traditional forms of empiricism, namely that observational foundations have a theoretical dimension. There is an analogue of this that also applies to inquiry which, we can say, is always theoretically motivated. Consider, for example, the task of opening a door by first unlocking it. Inquiry by means of trying out successive keys located on a large key ring is necessary because the door is locked, and has point because the inquiry strategy will terminate successfully when a key is found that opens the door.

The general point is that like a lock, all objects of inquiry come with some theoretical structure. If this structure is sufficiently rich to determine what is to count as a solution, or a successful end to inquiry, Plato's paradox can be resolved in an epistemically progressive way. Some problems, or research issues, exhibit a very high degree of solution specificity, so the degree of fit (or coherence) among the elements of theorized issue, methodology, and theory-laden observational evidence is quite compelling. However, the point does not lapse for want of well structured problems. Epistemic progress can occur in more incremental ways, where the degree of fit may be defined over less ambitious elements, such as better formulations of a research problem, or ways of partitioning the problem into more manageable components.

A less abstract way of stating matters is in terms of a constraint satisfaction view of inquiry. We appeal to the theoretical structure presumed in any particular inquiry to specify constraints on what counts as an answer to a research question, or problem. Although, as Haig (1987, p. 30) notes,

> … our significant research problems will not be fully structured and, therefore, will not constitute complete descriptions of their solutions … yet we articulate our problems in terms of their constituent constraints, and these constraints do serve to direct us towards their problems' respective solutions.

Moreover, this process is iterative, involving over time the successive application of satisfying multiple soft constraints in as coherent a manner as possible and feeding back proposed solutions into the theoretical machinery of further research issue and problem specification.

*Coherentist justification*

While constraint satisfaction accounts of research are well known in the literature (e.g. Nickles 1980; Haig 1987; Robinson 1993, 1998), there is a need to draw links between situated and specific contexts of successful constraint satisfaction implicit in the usual attempts to meet Plato's paradox, and the broader enterprise of attempting to articulate a post-empiricist account of knowledge and its justification, given the success of criticisms of epistemic foundationalism. This can be done by specifying more fully the epistemic notion of coherence that lies behind the normative adjudication of fit among the elements of inquiry.

To illustrate how coherence can operate, consider again the quest for foundations to knowledge. This was always theoretically motivated, assuming some view of the powers of the mind and how knowledge might be acquired. Mostly, these views were thought to be obvious commonsense, embodying folk-theoretical assumptions about the operation of sensory organs and the processing of sensory information. However, where these assumptions are less warranted than the foundations they demarcate are supposed to be, the point of foundational patterns of justification lapses. In building an epistemology it is more reasonable to make use of our most sophisticated scientific theories of how people acquire and process knowledge. But these theories occur fairly late in the build-up of knowledge, and certainly well after any of the usual candidates for privileged empirical foundations.

The lesson to draw from conceding the relevance of such disciplines as psychology, physiology, neuroscience and cognitive science to epistemology, is to abandon the demarcation task altogether and settle for developing the most coherent account of what we claim to know. Given that empirical evidence is never sufficient for theory choice, but that epistemically progressive choices are made all the time as we navigate our way through the environment, it would be more fruitful to attempt to discern how these choices are actually made.

Within any proposed framework for improving knowledge and enhancing inquiry, for doing better than chance or coin tossing, the following broad epistemic constraints are arguably central (see Quine and Ullian 1978; Evers and Lakomski 1991, pp. 37–45):

(a) *Empirical Adequacy:* There is a premium on theory driven empirical expectations being met, on predictions being fulfilled. This is just basic for navigating successful trajectories through the multiplicity of options presented by experience. Theories are therefore to be preferred over rivals to the extent that they achieve empirical adequacy.

(b) *Consistency:* As networks of claims that contain an inconsistency formally permit the deduction of any claim whatsoever, theory identity requires consistency. Without consistency, it is not possible to say where one theory ends and another begins. Nor is it possible to identify descriptions of particular states of affairs as being implied by the theory, thus compromising the nature of evidence for empirical adequacy.

(c) *Comprehensiveness:* Where explanation is seen as an advantage, theories that have the resources to explain more things rather than fewer are to be preferred.

(d) *Simplicity:* Theories that can explain the same phenomena while invoking fewer unexplained assumptions are to be preferred. This requirement penalizes recourse to *ad hoc* explanation — the practice of turning the inexplicable into a primitive additional assumption, or axiom, of current theory.

(e) *Learnability:* As a constraint on good theory development, learnability is aimed at ensuring that the account we give of known phenomena leaves the phenomena plausibly knowable. Thus, if we seek to explain human ethical judgment in terms of an abstract realm of ethical properties known only by intuition, and then fail to provide any account of the workings of this intuition, we offend against fulfilling the learnability requirement. It is a demand designed to ensure that accounts we give of things we know do not entail that they are unknowable.

(f) *Explanatory Unity:* This is really the result of combining both simplicity and comprehensiveness, in order to emphasize the advantages of being able to account for the most phenomena using the least theoretical resources.

Collectively, these criteria of theory choice are known as coherence criteria. It is claimed here that they constitute a set of soft constraints useful for guiding the development, or growth, of knowledge. Three sorts of arguments are relevant to defending their usefulness in providing a plausible distinction between justified and unjustified knowledge. First, arguments for various epistemological positions, whether foundational, paradigms views, skepticism, or postmodernist alternatives, in order to be persuasive characteristically employ strategies that conform to the theoretical virtues of coherence, especially if we omit the observational requirement of empirical adequacy and ignore the learnability requirement. To this extent coherence conditions, or a more limited subset, function as touchstone, or common theory, across otherwise divergent theoretical perspectives. Those who wish to argue against this finding will still end up attempting to conform to coherentist conditions.

Now if we add in the observational requirement, in as theoretical a sense as would satisfy any critic of traditional empiricism, with an eye to meeting the demands of successful navigation, then secondly, we can defend representationalism by the following consideration due to BonJour (1985, p. 171):

> If a system of beliefs remains coherent (and stable) over the long run while continuing to satisfy the Observation Requirement, then it is highly likely that there is some explanation (other than mere chance) for this fact … [the best explanation for this fact] … is that (a) the … beliefs which are claimed within the system, to be reliable are systematically caused by the sorts of situations which are depicted by their content, and (b) the entire system of beliefs corresponds, within a reasonable degree of approximation, to the independent reality which it purports to describe …

In our earlier discussion of the orthogonal paradigms of multi-perspectivism, it was noted that the perspectives one can adopt concerning some phenomenon will, after all, need to articulate at some point if they are isomorphic with respect to the background causal structure embedded in that phenomenon. That is, because narrative accounts of teacher conduct (for example) will be isomorphic in causally significant ways to physiological and behavioral accounts, positing the existence of some shared basis for theoretical articulation has more antecedent plausibility than positing a miraculous coincidence. Now the same point can be made in relation to BonJour's consideration. The reason it is not a miracle that a coherent empirically adequate theory can satisfy the observation requirement over the long run is because the theory in some important sense can be said to *represent* features of the world that give rise to what is observed.

Third, the study of natural learning systems — creatures with nervous systems that learn from their environment — suggests that learning accomplished by massively parallel neuronal architectures operates in terms of a network relaxing into the coherent satisfaction of multiple soft constraints (Rumelhart and McClelland 1986). It is a reasonable part of the naturalist's research agenda to expect an adequate characterization of coherent constraint satisfaction in neural networks (and their biologically realistic artificial models) to provide insights into the physical detail behind the operation of coherence criteria of theory choice defined over symbolic theory formulations (see Thagard 1989). The methodological point to be made is that we should use the sheer ubiquity of successful, piecemeal, epistemic practice that occurs throughout the natural world as a starting point for constructing guidelines for our own epistemic practices. It is more advantageous to assume that across the phylo-genetic continuum, nature has regularly produced creatures that do better than chance in solving their navigation problems, and learn from those solutions.

### The structure of educational theories

Just as logical empiricism's testability criteria of justification have consequences for structuring theories — namely by favoring hypothetico-deductive hierarchical statement structures as illustrated in Figure 9.2 — so coherentism's criteria, and a resulting concern to preserve natural representation, also have consequences. The general idea is to see inquiry from the perspective of an 'epistemic engine', namely, something that builds up an internal representation of patterns extracted from experience subject to the constraint of coherently meeting the observation requirement over the long run. Figure 9.3 captures this dynamical process schematically.

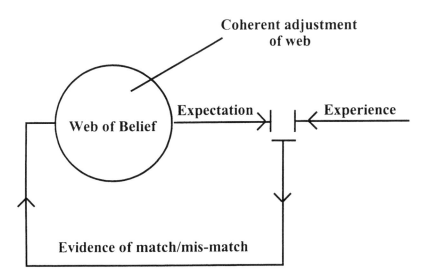

**Figure 9.3** Justification as epistemically progressive learning.

The important question concerns the kind of knowledge structure that develops in what has been labeled the 'Web of Belief'. For *sentential* representations of knowledge, Quine's elaboration of the metaphor is appropriate. It is a body of sentences, variously associated, with those more central to the organization of ideas and concepts in the web located at the centre, and those more directly in contact with sensory experience at the periphery. It is helpful to see the web structures in terms of the revisability of statements in the light of experience. At the centre of the web would be areas of knowledge such as logic, mathematics, and some branches of science, such as theoretical physics, where singular items of experience will have least impact because the intellectual costs of revision are so great and ramify across the web so extensively. Revisions at this level would need to be driven by invoking consideration of additional virtues of coherence. Statements at the periphery, however, singular observation reports, for example, can be revised quite readily in light of contrary experience, with no systematic consequences (Quine 1951; Quine and Ullian 1978). Note that, methodologically, even statements at the periphery can be rendered immune from revision if we are willing to make drastic adjustments elsewhere in the web (Quine 1951, pp. 40–43). In this respect, coherence criteria function as multiple soft constraints.

One consequence of this more relaxed approach to theory structure is a blurring of the distinction between natural and social science. This point can be made more strikingly by invoking naturalistic information-theoretic ideas to describe theories (For an introductory account see Chaitin 1975; Dennett 1991; Evers and Lakomski 1996, pp. 134–136). Consider a set of data points, individuated under some suitable schema for description. Call the total set of descriptions a 'bit map' of the data set. Then we can regard the data set as patterned, or non-random, if there exists a description of it that requires fewer bits of information than the bit map. Non-random data will therefore be *compressible*. For the relatively context invariant data sets of physics, or the level at which we are interested in phenomena conforming to the laws of arithmetic, extremely high levels of compression can be achieved, as is evident in the very simple equations that can be used to compress arbitrarily large bit map accounts of empirical data. But where context matters, compression grades off, not suddenly, but on a continuum. However, while the mainly linguistic theory formulations characteristic of the social sciences are clearly not as compressed as some physical science theories, they obviously comprehend significant regularities within contexts that are of great use in social navigation.

From this perspective not only are natural and social science on the same representational continuum, but in terms of coherentist justification they are also on the same methodological continuum. The so called qualitative/quantitative distinction is largely an artifact of disputes concerning foundational epistemologies and their models of theory structure.

## Towards the Non-Linguistic

The models of theory formulation that have so far been considered have been linguistic, or quasi-linguistic structures — hardly surprising since theorizing is a social activity and language is our prime mode of communication. The requirements of public communication should not, however, be translated uncritically into constraints on ways of representing the inner dynamics of knowledge and cognition. To do so is to end up seeing thought as a language-like, rule based, process of manipulating symbolic structures.

Unfortunately for this model, as we have been arguing, very little of the vast bulk of human cognitive activity can be captured, or formulated in this fashion. Most human cognitive activity is manifested in the form of skilled performance involving finely graded judgments influenced by multiple shifting constraints, as has correctly been noted and documented by the various cognitive anthropological studies in the situated learning/cognition mode to which we referred in earlier chapters. Linguistically expressed compression algorithms come nowhere near to representing an agent's practical knowledge, or to providing sound clues as to how such knowledge is acquired. The result is a familiar, but regrettable, filtering out of knowledge of practice from research, and the virtual institutionalizing of a methodological theory/practice divide. From an information-theoretic perspective, however, the actual coding of knowledge is not a basic issue. Earlier conclusions about representation, justification and the dynamics of knowledge acquisition can be expected to hold up across divergent codings. We turn to some of the implications of naturalistic representations of knowledge, for the researching of practice, in the next chapter.

# References

BonJour L. (1985). *The Structure of Empirical Knowledge*. (Cambridge, MA: Harvard University Press).

Chaitin G. (1975). Randomness and mathematical proof, *Scientific American*, **232**(5), pp. 47–52.

Connelly F.M. and Clandinin D.J. (1990). Stories of experience and narrative in inquiry, *Educational Researcher*, **19**(5), pp. 2–14.

Clandinin D.J. and Connelly F.M. (1996). Teachers' professional knowledge landscapes: Teacher stories — stories of teachers — school stories — stories of schools, *Educational Researcher*, **25**(3), pp. 24–30.

Constas M. (1998). The changing nature of educational research and a critique of postmodernism, *Educational Researcher*, **27**(2), pp. 26–33.

Cussins A. (1992). The limitations of pluralism, in D. Charles and K. Lennon (eds.) *Reduction, Explanation, and Realism*. (Oxford: Clarendon Press).

Dennett D. (1991). Real patterns, *Journal of Philosophy*, **88**(1), pp. 27–51.

Evers C.W. (1991). Towards a coherentist theory of validity, *International Journal of Educational Research*, **15**(6), pp. 521–535.

Evers C.W. (1997). Philosophy of education: a naturalistic perspective, in D. N. Aspin (ed.) *Logical Empiricism and Post-Empiricism in Educational Discourse*. (Durban: Heinemann).

Evers C.W. (1998). Decision-making, models of mind and the new cognitive science, *Journal of School Leadership*, **8**(2), pp. 94–108.

Evers C.W. and Lakomski G. (1991). *Knowing Educational Administration*. (Oxford: Pergamon Press).

Evers C.W. and Lakomski G. (1996). *Exploring Educational Administration*. (Oxford: Pergamon Press).

Haig B.D. (1987). Scientific problems and the conduct of research, *Educational Philosophy and Theory*, **19**(2), pp. 22–32.

Husén T. (1997). Research paradigms in education, in J. P. Keeves (ed.) *Educational Research, Methodology and Measurement: An International Handbook*. (Oxford: Pergamon Press, 2nd Edition).

Kerlinger F. (1986). *Foundations of Behavioral Research*. (New York: Harcourt Brace, 3rd Edition).

Kuhn T. (1970). *The Structure of Scientific Revolutions*. (Chicago: University of Chicago Press, 2nd Edition, Enlarged).

Linsky B. (1971). (ed.) *Reference and Modality*. (London: Oxford University Press).

McMillan J.H. and Schumacher S. (1993). *Research in Education*: *A Conceptual Introduction*. (New York: Harper Collins, 3rd Edition).

Nickles T. (1980). Scientific problems: three empiricist models, in R. Giere and P. Asquith (eds.) *Philosophy of Science Association 1980*, Volume 1. (East Lansing: Philosophy of Science Association).

Petrie H. (1972). Theories are tested by observing the facts: or are they?, in L.G. Thomas (ed.) *Philosophical Redirection of Educational Research*, 71st NSSE Yearbook. (Illinois: University of Chicago Press).

Quine W.V. (1951). Two dogmas of empiricism, *Philosophical Review*, **60**, pp. 20–43.

Quine W.V. and Ullian J.S. (1978). *The Web of Belief*. (New York: Random House, 2nd Edition).

Robinson V.M.J. (1993). *Problem-Based Methodology*. (Oxford: Pergamon Press).

Robinson V.M.J. (1998). Methodology and the research-practice gap, *Educational Researcher*, **27**(1), pp. 17–26.

Rumelhart D.E. and McClelland J.L. (eds.) (1986). *Parallel Distributed Processing*, Volume 1. (Cambridge, MA: MIT Press).

Thagard P. (1989). Explanatory coherence, *Behavioral and Brain Sciences*, **12**, pp. 435–467.

# 10

# Researching Practice

Developments of modes of representing knowledge and thought that take as their model the parallel processing architectures of the brain, have occurred mainly in engineering and computer science, and more recently in cognitive science. In engineering and computer science, where the demands for good, efficient, real time computation performing software for pattern recognition tasks are legion, there is little incentive to constrain software design by fidelity to what is biologically realistic. The main demand is the application, or task, at hand. In the loose usage that prevails in this respect in cognitive science, the term 'connectionism' covers these kinds of developments. In psychology, and much cognitive science, where there is great interest in constructing models that capture significant features of human information processing, some effort, reflected in the use of the term 'neural networks', is made towards model realism. By virtue of the fact that science often proceeds by a process of initial simplification followed by a build up of complexity, the practices of engineers and cognitive scientists, in this field, are best located on a continuum, rather than being seen as entirely different.

Where there is a gap, at least in terms of the amount of research available, it is in the extension of ideas from the new cognitive science into mainstream social science. A very impressive body of research exists in philosophy, centred on exploring the implications of neural networks for understanding epistemology, knowledge representation, thought, explanation, decision-making, ethics, and semantics. Some of this work we have already discussed, particularly the writings of Andy Clark (1989, 1993, 1997), Paul and Patricia Churchland, (Churchland P.M. 1989, 1995, 1996, 1998; Churchland P.S. 1986; Churchland and Sejnowski 1990, 1994) and Paul Thagard (1996, 1998; Thagard and Millgram 1995; Thagard and Verbeurgt 1998). There is also a considerable amount of research being done on providing neural network models of natural language and its acquisition. (For an up to date account, see Elman *et al.* 1997.) For the study of social processes, however, less has been written. Key, systematic, works that we have earlier referred to are Hutchins's (1995) masterful *Cognition in the Wild* and Strauss and Quinn's (1997) pioneering *A Cognitive Theory of Cultural Meaning*. Both these books offer major insights into social and cultural phenomena through their various articulations with neural network accounts of individual and group cognitive activity.

New approaches to knowledge representation (and justification) can influence substantive bodies of theory in a number of ways. In the preceding chapters, we have tried to focus on the task of displaying influence through *reconceptualization*. In a series of studies, we have sought to show that knowledge viewed from the perspective of how the brain might learn and know holds the promise of yielding a better understanding of non-symbolic learning and skilled performance. Indeed, we have gone further, suggesting that the familiar linguistic forms of public communication are subsumed as patterns subject to identical modes of processing.

Another source of influence, and a further development from reconceptualization, is to be found in the extraction of data and their analysis. That is, we should be able to use ideas drawn from neural network modeling as a research tool for exploring and capturing the practical regularities to be found everywhere in the social world. Because artificial neural networks are learning devices which, for the most common varieties in use, attempt to build up a sub-symbolic representation of the statistical structure of data sets through a process of trial and error (subject to the demand to minimize error which is defined as the difference between the network's output and some target output, or maximizing 'harmony', which is the case with, say, Thagard's constraint satisfaction networks) they can be useful in research for *discovering* patterns as well as for theorizing about the nature of pattern in social science and its representation. For after successful learning has taken place, the array of numbers that constitutes the connection strengths, or weights, between layers of artificial neurons comprising the net, and the characteristic resulting activation patterns in those layers, is where the sub-symbolic representation of knowledge occurs. This is the compression algorithm summarizing the regularities the net has extracted from the data comprising its experience. In what follows, we first review, for research purposes, some central features of the reconceptualizations we commend. We then examine briefly a theoretical issue arising out of the use of nets for data extraction and analysis.

## Reconceptualizing the Tools of Inquiry

In theorizing about actions and practices in a variety of contexts, we want to single out four conceptual tools for dealing with questions of evaluation, for locating the assessment of practices within the broader framework of justification and evaluation that prevails for symbolically represented knowledge.

### Explanation and prototypes

In the course of a busy day, an administrator will initiate many actions that reflect some kind of unconscious, inexplicit, theorizing. A teacher sounds unwell, over the telephone, in reporting in sick, and the person responding makes a judgement that an emergency replacement will be needed for more than one day. A piece of student misbehavior is judged to reflect a troubled person rather than a troublesome person.

A student with special needs is admitted to a school's facility for teaching hearing impaired students. This will stretch the student staff ratio a bit beyond five to one, but the student has a good record and presents as eager and cooperative. A judgement is made to admit. A teacher is having trouble with a difficult class, but the situation is sensed to require encouragement and support rather than the usual speech about classroom management. The computer studies faculty put in yet another request for additional funding which needs to be met by making reductions elsewhere. Here, a complex judgement will have to be made that involves juggling considerations of equity and fairness of budget allocations with educational priorities and goals. And so on.

In researching these practices, our point is that eliciting verbal responses from administrators who are being prompted to explain their decisions, judgements, and actions, imposes a mode of representing practical know-how that can clearly fail to capture the structure and detail of administrators' real knowledge. Questioning in this mode, implies a hypothetico-deductive model of explanation, where a person's action description is constrained by an expectation that it be an inference from some more general principle, or set of principles, comprising the person's (linguistically formulated) theory. As we have seen, however, these actions may best be theorized as being the result of activating learned prototypes. (See Churchland 1989, pp. 197–230; 1995, pp. 97–121.) In the above examples, prototypes acquired through experience might include the typical sound of someone talking with a heavy cold, types of misbehaviour associated with particular known student background conditions, prototypical modes of self presentation for students with different special needs, varieties of difficult classes, and a range of ethical prototypes that are endlessly invoked and refined in almost all dealings with people. In all cases, what counts as an explanation of the particular action is whether it falls under some prototype. Roughly speaking, a skilled person is acting the way they are because this is the kind of situation that calls for that kind of action.

We should also observe that senior administrators are often involved in deciding on atypical problems or issues, where only extensive and varied experience will provide the opportunity to develop an appropriate prototype. Here, adjustments to a person's physically instantiated web of belief may be additionally generated by symbolic patterns — by reading, or listening to the experiences of others. Conceptual revisions can occur which over-ride familiar perception based similarities; whales classified under the 'fish' prototype can come to be seen as more similar to kangaroos when classified under the more theoretically parsimonious prototype of mammals.

An interesting question that flows from this characterization of research focus is the matter of how prototypical activation patterns can be identified, or recorded, as research evidence of learned expertise. We say something about this issue in the discussion below on data acquisition and analysis.

*Inference to the best explanation*

Inference to the best explanation is already a coherentist notion, with most elaborations in the methodology literature being given in terms of features, or properties, of symbolic structures. (See a number of the essays in Boyd , Gasper and Trout 1991.) When it comes to practice we have seen, in Chapter 6, that a decision to do something, in fact a resulting action, can be reconceptualized as an inference to the best action flowing from an attempt to maximize the satisfaction of multiple soft constraints (Thagard and Millgram 1995). That is, inference to the best explanation, construed in relation to practice, *is* action that is an outcome of a coherentist constraint satisfaction process that admits of some preliminary neural modeling.

More generally, we should view intelligent action as a relatively global phenomenon, to be treated in terms of a holistic evaluation of multiply competing goals and alternative possible actions. Action outcomes are to be seen within the context of coherent systematic plans reconstrued as sets of features (goals and actions) linked together by positive constraints. And competing systematic plans — all linked by positive *and* negative constraints — are in turn evaluated by the application of a suitable constraint satisfaction model.

*Ways of living*

Applying coherentist epistemology to the matter of theory choice, because of its global nature, goes over into the context of non-symbolic practice as determining and evaluating a way of living. As such, it is an application of modeling practical knowledge that extends the search for coherent plans across the entire spectrum of planned action, seeking the maximization of constraint satisfaction among all features. While the maximum satisfaction of both positive and negative constraints may partition goals and courses of action into smaller clusters, we would expect sufficient integrity of linking to sustain the identification of cultural practices. This is a reasonable expectation given that the learning of culture is mostly shaped by cultural factors to do with the history of one's learning trajectory.

Strauss and Quinn's (1997) attempt to give a connectionist account of culture provides a powerful elaboration of these ideas. Beginning with a 'cognitive' view of culture, they have this to say about its meaning:

> To the extent people have recurring, common experiences — experiences mediated by humanly created products and learned practices that lead them to develop a set of similar schemas — it makes sense to say they share a culture. … [I]t consists of regular occurrences in the humanly created world, in the schemas people share as a result of these, and in the interactions between these schemas and this world. (Strauss and Quinn 1997, p. 7)

Although formulated in the language of schemas, they endorse a thoroughly connectionist reduction of this term of explanatorily high level psychological theory, a reduction that coheres well with our own emphasis on constraint satisfaction. Take, for example, the culture-laden understanding of the schema for teacher. On one understanding, we might expect it to involve a number of associated features, say, educated, works in a school, completed a Bachelor of Education degree, skilled speaker, likes children, and so on. Call each of these features 'knowledge atoms'. (Smolensky 1986, pp. 204–208.) Each of these 'atoms' is, in turn, represented by a sequence of network nodes whose levels of activation, for that atom, could then be described in the form of a vector. The schema for teacher is thus to be thought of as a coherent assembly of knowledge atoms (Smolensky 1986, p. 203). However, the principle of coherence, at the network level, is a dynamical process of constraint satisfaction. Some 'atoms' will be activated in the presence of some inputs and not others, as the net 'settles' into a stable 'interpretation' of its input data. If the same principles for identifying knowledge atoms with sets of nodes is employed, prototypes become instances of schemata, whose physical basis for activation is the pattern and value of all the weighted signals being transmitted from the previous network layer.

Strauss and Quinn (1997, pp. 61–76) describe a connectionist model for extracting patterns of regularity reflecting appropriate modes of address (e.g. Mr., Ms., Uncle, pet name, last name, first name, etc.) used by different people in the US, at different ages. These usages distil complex cultural patterns, having age, gender, class, ethnic, and regional influences. The reconceptualization of culture in terms of schemata, and schemata in terms of the operation of neural networks, coheres well with the theorizing of much cultural knowledge as non-symbolic and practice oriented, though with a focus on very broad categories of cognitive similarity emerging from the satisfaction of what amount to cultural constraints on learning. This is a conceptual tool that can, of course, be applied to the understanding of organizational culture and its dynamics.

*The extended mind*

The last major conceptual tool we wish to mention is conceiving practice within the framework of the extended mind. In earlier chapters, we have highlighted two respects in which it is fruitful to see the mind as extending beyond the boundary of an individual skull. The first applies to groups, or teams, or organizations, or societies. Analogous to the way an individual's knowledge is distributed across linked assemblies of neurons, for some conceptual purposes it is of value to locate an organization's knowledge in its distributed, yet coordinated, patterns of activity among its many individuals. In Hutchins's (1995) work this has led to insights about organizational learning being generated by constraint satisfaction models of neural network learning. As many of the phenomena of administrative life are group, or organizational matters, such modeling can be a powerful aid to organizational research.

The second way in which individual minds can be extended is through artifacts, especially those which serve the explicit function of permitting cognitive 'off-loading' — external aids to memory and calculation and thought, such as calculators, books, pen and paper, writing, language, computers. Artifacts are not only bearers of cognitively significant patterns, they can also assist in pattern processing. These are vital factors in being able to give a full analysis of educational practices, which involve extensive interaction with educational artifacts. It is useful to see organizational designs as attenuated artifacts that mesh with the artifacts of building architecture, office plans and layouts, and the physical accessibility of avenues for communication. Hutchins's work on the design of aircraft cockpits, or the increasing consideration being given to the design of machines, such as Automatic Telling Machines, with which humans interact, displays the importance of the artifactual in an understanding of practice.

## Data Extraction and Analysis

In concluding this brief survey of the research implications of our naturalistic approach to theorizing practice, we mention one major theoretical insight as to the real representational powers of neural networks.

There is no question that connectionist models are powerful devices for the extraction of patterns among data. However, for our purposes, the key issue is whether information about what nets learn can be identified in ways that map onto the theoretical categories we have been trading in to develop a more fine grained causal account of practical knowledge representation. The good news is that many ANN software packages (e.g. the program Tlearn in Plunkett and Elman 1997) contain features that display the net's data configurations in the required ways. Tlearn, for example, displays characteristic activation patterns for learned prototypes, configurations of numerical weight values — architecture plus numbers — and, perhaps most useful of all, cluster analyses in the form of diagrams called dendograms that display the learned statistical structure of relations among input data items.

What relation do these displays of data values and architecture have to the patterns identified by human brains? The key theoretical result here, and an amazing one, is that the sorts of patterns detected in inputted data, for example as manifested in cluster analysis, or displayed in dendograms, are mostly stable across different network designs, drawing their properties primarily from the data (Churchland 1998). That is, inasmuch as a network has the resources to detect a pattern, the pattern is relatively invariant with respect to the pattern extractor. One consequence is that concepts identified as prototypes can be regarded as conceptually similar in different nets. We have a criterion of conceptual similarity that does not depend on the particular nature of the net and its fine detail. The concepts are in the data, as it were, not in the network. This means that limited, artificial models, of brain cognitive processes have a greater likelihood of capturing significant features of the structure of human practical knowledge, built up from experience, than might have been supposed. The big difference between even the most realistically designed artificial networks and the real network is assuredly one of huge complexity and processing resources. Nevertheless, it now seems reasonable to expect a structured isomorphism to obtain between the information physically instantiated in the brain and the information structures in the same data sets as revealed by more modest ANN processing resources.

Inquiry, construed both broadly and modestly, as that repertoire of theorized strategies, techniques, and procedures which enables us to act in the world with a realistic expectation of doing better than chance, can be informed by a range of sources. It is one of the strengths of combining naturalism with coherentism that it is able to acknowledge the successes of the many natural and commonplace forms of inquiry all around while being willing to posit a wider variety of factors that make for that epistemic success. To the extent that coherentism is able to draw on the resources of natural science, particularly recent developments in cognitive science, is indicative of one of its own epistemic virtues — namely that accounts of inquiry should be able to profit from their own recursive application.

The overall gain to be had from a focus on naturalism in research is an explanation of human thought and action which coheres with what is known about humans as biological beings who negotiate their complexly structured environments, mostly with success, often with great skill, and sometimes with finesse. While we do not yet possess the means appropriate for representing our skilled performance that are comparable with those developed for symbolically construed performance, we at least begin to understand how it is possible that a child learns to ride a bicycle, and what makes a person the skilled practitioner they are. Many more avenues need to be explored about how our minds/brains work, but in the face of such complexity we can take heart from Andy Clark's (1997, p. 175) delightful phrase that '... if *the brain* were so simple that a single approach could unlock its secrets, *we* would be so simple that we couldn't do the job! '

# References

Boyd R., Gasper P. and Trout J. D. (eds.) (1991). *The Philosophy of Science*. (Cambridge, MA: MIT Press).

Churchland P.M. (1989). *A Neurocomputational Perspective*. (Cambridge, MA: MIT Press).

Churchland P.M. (1995). *The Engine of Reason, The Seat of the Soul*. (Cambridge, MA: MIT Press).

Churchland P.M. (1996). The neural representation of the social world, in L. May, M. Friedman, and A. Clark (eds.) *Mind and Morals*. (Cambridge, MA: MIT Press).

Churchland P.M. (1998). Conceptual similarity across sensory and neural diversity: the Fodor/Lepore challenge answered, *Journal of Philosophy*, 95(1), pp. 5–32.

Churchland P.S. (1986). *Neurophilosophy: Towards a Unified Science of the Mind/Brain*. (Cambridge, MA: MIT Press).

Churchland P.S. and Sejnowski T.J. (1990). Neural representation and neural computation, in W. G. Lycan (ed.) *Mind and Cognition — A Reader*. (Oxford: Basil Blackwell).

Churchland P.S. and Sejnowski T.J. (1994). *The Computational Brain*. (Cambridge, MA: MIT Press).

Clark A. (1989). *Microcognition: Philosophy, Cognitive science, and Parallel Distributed Processing*. (Cambridge, MA: MIT Press).

Clark A. (1993). *Associative Engines: Connectionism, Concepts, and Representational Change*. (Cambridge, MA: MIT Press).

Clark A. (1997). *Being There: Putting Brain, Body, and World Together Again*. (Cambridge, MA: MIT Press).

Elman J.L. *et al.* (1997). *Rethinking Innateness: A Connectionist Perspective on Development*. (Cambridge, MA: MIT Press).

Hutchins E. (1991). The social organization of distributed cognition, in L. B. Resnick, J. L. Levine and S. D. Teasley (eds.) *Perspectives on Socially Shared Cognition*. (Washington, DC: American Psychological Association).

Hutchins E. (1995). *Cognition in the Wild*. (Cambridge, MA: MIT Press).

May L., Friedman M and Clark A. (eds.) *Mind and Morals*. (Cambridge, MA: MIT Press).

Plunkett K. and Elman J.L. (1997). *Exercises in Rethinking Innateness: A Handbook for Connectionist Simulations*. (Cambridge, MA: MIT Press).

Smolensky P. (1986). Information processing in dynamical systems: foundations of harmony theory, in Rumelhart D. E. and McClelland J. L. (eds.) *Parallel Distributed Processing*, Volume 1. (Cambridge, MA: MIT Press).

Strauss C. and Quinn N. (1997). *A Cognitive Theory of Cultural Meaning*. (Cambridge: Cambridge University Press).

Thagard P. (1996). *Mind: An Introduction to Cognitive Science*. (Cambridge, MA: MIT Press).

Thagard P. (1998). Ethical coherence, *Philosophical Psychology*, **11**(4), pp. 405–422.

Thagard P. and Millgram E. (1995). Inference to the best plan: A coherence theory of decision, in Ram A. and Leake D.B. (eds.) *Goal-Driven Learning*. (Cambridge, Mass.: M.I.T. Press), pp. 439–454.

Thagard P. and Verbeurgt K. (1998). Coherence as constraint satisfaction, *Cognitive Science*, **22**(1), pp. 1–24.

# Author Index

# Subject Index